From the Lorentz Transformation to the Dirac Equation

A Whirlwind Tour of Special Relativity

K. Thomas R. Davies

Daniel S. Nydick

ISBN: 1517772176
ISBN 13: 978-1517772178

Contents

Preface

The material for this textbook on Special Relativity was developed over a period of about sixteen years in various courses at Duquesne University taught by one of us (KTRD). The level of this textbook is aimed at upper-level undergraduate students or first-year graduate students and presupposes a knowledge of basic physics and differential and integral calculus. During the period when the Duquesne courses were taught, there was not a single text available that contained all of the material covered in this text or contained all of the various approaches covered in the text. However, the material covered in this text can be found in various books, all of which are listed in the Bibliography, and we would claim that this text offers a unique perspective to Special Relativity. Moreover, while the primary emphasis in this book is concerned with Special Relativity, we do devote all of Chapter 11 and part of Chapter 13 to General Relativity (just to whet the reader's appetite for additional relativity).

The structure of this book is as follows: The Introduction in Chapter 1 gives pertinent background of Special Relativity, which is rooted in the history of electricity and magnetism, paricularly in the nineteenth century. Then, in Chapter 2, we present the Lorentz equations, a rigorous derivation of which is given in Appendix A. The Lorentz equations are essential to understanding the theory of and phenomena related to Special Relativity since the results presented in Chapters 2–10 follow in a straightforward manner from these important equations. For example, three significant applications are discussed in Chapter 3—namely, simultaneity (or the lack thereof), time dilation, and the Lorentz contraction.

In Chapter 4 a number of important concepts are discussed: four vectors, including differential four vectors and operations with four vectors; time-like, space-like, and null four vectors and trajectories; proper length and proper time; light cones and world lines; and causality. Causality dictates that, while time travel into the future is in principle possible, time travel into the past is strictly forbidden. One of the most useful results presented in this chapter is that the four-dimensional dot product of two four vectors is an invariant; i.e., it will be the same in any relativistic reference frame (a so-called inertial frame). The formalism of four vectors is continued in Chapter 5, in which we develop the kinematics and dynamics of Special Relativity, including the famous equation: $E = mc^2$. Comparisons are made too with Newtonian kinematics and dynamics.

Chapter 6 presents the variational principle that is useful in both Special Relativity and General Relativity. This follows from the calculus of variations and is known as "The Principle of Extremal Aging". An important result is that, in Special Relativity, the relativistic energy and vector momentum are separately conserved in any inertial frame. Chapter 7 discusses the applications of the relativistic equations to the threshold scattering problem and other scattering or decay problems encountered in elementary-particle physics. For the threshold problem, a useful result is that there is an important invariant quantity that is conserved in either the center-of-mass or laboratory reference frames.

Chapter 8 details the relativistic Doppler effect. It is shown that the effect is the same whether the source or observer is moving with the same relative velocity. We discuss the situation for the colinear configuration of source and observer and also when the source and observer are at an angle with respect to one another. The nonrelativistic limit is considered. Important applications involve Hubble's Law and relativistic beaming. We mention too that from Hubble's Law, one concludes that beyond a certain distance in space (the Hubble distance), the velocity of recession of distant galaxies exceeds the speed of light. This superluminal motion, other types of which will be explored in Chapter 10, is a legitimate effect arising from the expansion of space allowed in General Relativity.

In Chapter 9, we show that the Maxwell equations, involving the $\vec{E}$ and $\vec{B}$ fields, satisfy covariant relations. In particular the vector and scalar potentials that give the correct solutions for the Maxwell equations are shown to satisfy d'Alembertian relations that are manifestly covariant, another result that follows directly from the four-dimensional dot-product invariance mentioned above. We emphasize too that this result is obtained using the so-called Lorentz gauge, which we are free to choose.

As mentioned above, other types of behavior related to superluminal motion are explored in Chapter 10. These examples are (i) Cherenkov radiation and the sonic boom; (ii) group and phase velocities; and (iii) the (apparent) superluminal expansion of clouds (or "blobs") from certain distant radio sources. As we will explain, this last case turns out to be an optical illusion, although the clouds are moving at speeds that are a sizable fraction of the speed of light.

In Chapter 11, we present a short summary of the most important ideas and results from Einstein's Theory of General Relativity, which pertains to gravity. Perhaps one of the most intriguing ideas is the notion that gravity is not a force in the usual sense since mass/energy determine SpaceTime geometry and, in turn, the curvature and geometry of SpaceTime determine the motion of matter. Thus, there is a kind of "self-consistency" between mass/energy and the SpaceTime geometry, an idea that is formally embodied in the famous Einstein equations. The trajectories followed by particles arise from the "Principle of Extremal Aging", mentioned above, and are called geodesics. Also, General Relativity solves the problem of "Instantaneous Action at a Distance", which always plagued Newtonian gravitatonal theory. Moreover, in this chapter, we discuss the Equivalence Principle, tidal forces, and the concept of "free fall". We next analyze the most important confirmed predictions of General Relativity, which are: (a) gravitational time dilation (or equivalently the gravitational red shift); (b) precession of the perihelion of the planet Mercury (and the other inner planets); (c) the bending of starlight passing near the Sun (or other massive body); (d) the slowing of radio waves passing near the Sun, from signals propagated between the Earth and Mars, Mecury, and Venus; (e) the existence of Black Holes; and (f) the Big Bang Theory of the evolution of the Universe.

Next, Chapter 12 gives a derivation of the famous Dirac equation, which is a union of Special Relativity and Quantum Mechanics. Also, there is a discussion of the intrinsic covariance of the Dirac equation.

We conclude the book with Chapter 13, which gives a short summary of the relativity of measurements in modern theories of physics; particularly measurements made in Quantum Mechanics, Special Relativity, and General Relativity.

Finally, we acknowledge useful discussions with S. Cohen, R. W. Davies, S. Frittelli, W. T. Griffith Jr., T. Maurer, and R. Shurig.

KTRD acknowedges the support of his wife Carol and dedicates this book to her.

Chapter **1**

Introduction

This is a book on Special Relativity (SR), on which we will mainly focus. However, the background of SR is intimately connected to the history of Electricity and Magnetism (E&M), which is detailed in the Special Appendix and which covers a period from Thales in ancient times to modern discoveries at the end of the twentieth century. This appendix is a research project in its own right and is important background for many of the chapters of this book.

1.1. Remarks about the Special Appendix

The Special Appendix details all of the fundamental discoveries and developments associated with E&M and/or photons, the "carriers" of the electromagnetic force. These include: (i) the theory of QED (Quantum Electrodynamics), which is the relativistic quantum mechanical theory of the *interaction of the electromagnetic force with charged particles*, and (ii) the ElectroWeak Theory, which is the *merging of QED with the weak force* and which also necessitates including in the general history various developments associated with the weak interactions (such as the theoretical prediction and experimental verification of the violation of parity in the mid-1950s). We also include in this Appendix discoveries associated, not only with visible light, but with other parts of the electromagnetic spectrum: radio waves, ultraviolet light, infrared light, x-rays, and gamma rays. This leads us to to some (but not all) discoveries from astronomy, astrophysics, and cosmology since our knowledge of our universe has depended upon the information that we obtain via measurements of distant objects using the electromagnetic spectrum. As mentioned too at the beginning of the Appendix, certain commercial or practical branches emanating from the general history of E&M are not included. Moreover, by its very nature, the Special Appendix covers many of the twentieth-century revolutions to be discussed below.

1.2. Twentieth-Century Physics Revolutions

The twentieth century witnessed the greatest explosion of physics ever seen. There were many revolutions encompassing a number of different fields of physics, some of them related to one another, and it included both theoretical and experimental contributions. Many of the developments fed off of one another, as the revolutions (several of which really began in the nineteenth century) continued their evolution. For example, toward the end of the twentieth century, the revolutions in elementary-particle physics dovetailed with and reinforced the almost simultaneous revolution in astrophysics/cosmology. The symbiosis of these two fields is interesting because it showed the intimate connection between the smallest particles known to man and the largest structures in the Universe. As noted by many authors, this connection can be seen most dramatically by considering the beginning of the Universe at the Big Bang, just after which the whole Universe itself consisted of a tiny region of quantum dimensions.

The early major twentieth-century breakthroughs were true paradigm shifts—first Albert Einstein's *Theory of Special Relativity (SR)* in 1905, which was followed by Einstein's *General Theory of Relativity (GR)* in 1916. Also, in 1905, Einstein formulated the theoretical basis for the photoelectric effect, which is a theory built upon the work of Max Planck (1900) and was the beginning of *Quantum Mechanics*, which was the third major early twentieth-century paradigm shift. Then, Quantum Mechanics was further sparked and advanced by the fundamental contributions of Niels Bohr in 1913, Louis deBroglie in 1924, Erwin Schrödinger in 1926, and Werner Heisenberg and Paul Dirac in 1927. Further discussions of these theories, and some of their experimental verifications, are given in the Special Appendix. (It is interesting too that Einstein was one of the founders of Quantum Mechanics. Yet, in later years he became an opponent of quantum theory, having many arguments with Bohr, because he objected to the probabilistic nature of the theory.)

At any rate, SR, GR, and Quantum Mechanics set the tone and were the basis for the remaining revolutions in physics that would occur in the century. For example, in section 1.1 above, we mentioned QED, which was one of the central revolutions in physics in the twentieth century and which extensively utilizes the union of both SR and quantum theory. It would take us too far afield to detail the remaining revolutions, which are also thoroughly discussed in the many references in the Bibliography of this book.

1.3. Historical Background of SR

For the background of SR, we shall be especially concerned with the history of E&M in the nineteenth century. There are two main developments that we shall address in the subsections to follow:

(1) The debate, over several centuries, concerning *whether light is a particle or a wave.*

This is important since in our and many other texts on SR, light is treated as both a wave and a particle. Thus, we should try to understand the general context of this behavior.

(2) The implications of the *ether* (or aether) theory of light developed in the nineteenth century.

Many of the most important unsolved pre-relativity problems were associated with the ether theory. The ether theory and related problems of classical physics provide the immediate backdrop in the nineteenth century, which led to the theory of SR, developed by Einstein in 1905.

1.3.1. Is Light a Particle or a Wave? The Particle-Wave Duality. We know that light exhibits wave properties. e.g., *diffraction* can only be easily explained by assuming a wavelike character for light. However, there has been a long history involving a famous controversy that questions whether light is truly a wave. In 1678, Christian Huygens' work on reflection and refraction challenged the prevailing view that light was "corpuscular", or particle-like. Unfortunately, in the late 1600's Sir Isaac Newton believed that light was not wavelike, as Huygens claimed. Some historians believe that Newton threw his support to the older corpuscular theory, which (under his great prestige) continued as the official theory. In any case, the wave theory of light was strongly resisted for more than another century. Then, in the early to mid-1800s, the work of Thomas Young (1801, 1817), Augustin-Jean Fresnel (1819), Armand Fizeau (1850), and Léon Foucault (1850) proved conclusively that light is indeed a transverse wave motion. See the Special Appendix for details of these and other topics discussed in this subsection.

The development of classical E&M moved with great speed in the 1800s (particularly with the work of André-Marie Ampère in 1820, Michael Faraday in 1821–1831, and Joseph Henry in 1829), culminating in the formulation of the four fundamental equations of the theory by James Clerk Maxwell (1873), who proved that light is an electromagnetic wave, thereby unifying E&M and optics. Maxwell's equations were partial differential equations involving the time- and space-dependent $\vec{E}$ and $\vec{B}$ fields. (See Chapter 9 of this book.) Alas, Maxwell's original presentation of his work was extremely difficult to comprehend. However, in the 1870s Oliver Heaviside, an unemployed telegrapher, made the four Maxwell's equations clear and understandable, expressing them in essentially the form that we know today.

Moreover, during the early- to mid-1800's, since *infrared radiation* (W. Herschel in 1800 and M. Melloni in 1850) and *ultraviolet radiation* (J. Ritter in 1801) were known to have properties similar to light, it became apparent that there is a spectrum of electromagnetic waves. Other parts of the spectrum were discovered at later dates (*radio waves* by Heinrich Hertz in 1886, *x rays* by Wilhelm Röntgen in 1895, and *gamma rays* by Paul Villard in 1900).

However, the story regarding the wave-particle controversy continues. In the early twentieth century, physicists developed the theory of Quantum Mechanics, which predicts that light is transmitted by particles called *photons*. This result has been verified in the photoelectric effect, which clearly demonstrates the particle-like nature of the photon and, therefore, of visible light and other parts of the electromagnetic spectrum. (See the Special Appendix for the research contributions of Philipp von Lenard in 1902, Albert Einstein in 1905, and Arthur Compton in 1923.) Nevertheless, the many experiments involving interference and related phenomena clearly verify that *light is also a wave*. Thus, we see that, since Newton's time, we have come full circle, but the fundamental question still remains: *is light a wave or a particle?*

On the other hand, quantum theory also dictates that *ordinary matter* at the microscopic level (e.g., fundamental particles) *exhibits both particle- and wave-like properties*. For example, in 1927 Clinton Davisson and Lerner Germer performed an experiment proving that electrons could be diffracted through a nickel crystal (also independently performed by G. P. Thomson), thereby establishing that electrons have a very short wavelength. This means that under some circumstances *matter also can behave like a wave*.

One dilemma of modern physics is that both light and matter exhibit properties of waves *and* of particles. In some sense, this problem is resolved by saying that what effect you observe depends very strongly upon how you ask your question and the way in which you perform your measurements. (I.e., you can do one type of experiment that measures particle-like properties and another, that measures wave-like properties. See also Chapter 13 of this book.) Most physicists have become comfortable with this dichotomy, seeing these twin properties as a fundamental "complementarity" of nature. It has been suggested that we refer to light and particles as *"wavicles"*, thereby accepting a permanent kind of coexistence. Thus, this *"particle-wave duality"* might imply that either waves or particles are only approximations to reality. These puzzles are obviously related to the *probabilistic*, or *nondeterministic*, nature of Quantum Mechanics.

Over the years, many philosophers and scientists have been very uncomfortable with the theory of Quantum Mechanics, particularly the features associated with its probabilistic nature. As mentioned above, Albert Einstein, one of the founders of Quantum Mechanics (because of his work on the photoelectric effect*), challenged the philosophical foundations of the theory. He had famous debates with Niels Bohr and insisted that "God does not throw dice." After Einstein's time, other scientists actively searched for alternative theories to Quantum Mechanics. Also, a number of experiments have been performed to check the validity of Quantum Mechanics, and so far the theory has passed all tests. However, in recent years a number of physicists have been drawn into the new and seductive field of quantum information theory (and related topics), which deals with the philosophy of Quantum Mechanics. Moreover, many working physicists simply use the rules of the theory to make calculations or perform experiments, without worrying very much about the deep, underlying philosophy of Quantum Mechanics. It should be mentioned too that almost all of the modern experts on Quantum Mechanics and/or Quantum Field Theory have taken the point of view that the only legitimate position is the wave interpretation of matter and radiation. For example, see Art Hobson's interesting paper "There are no particles, there are only fields" [73], which contains a number of additional references. Since all particles have fields, it would appear that the wave interpretation is the correct representation.

From the SR point of view, it is important to be able to treat light as both a wave and a particle. However, we shall not worry about the subtleties of Quantum Mechanics since SR is a "classical" (i.e., nonquantum) theory. As we shall indicate in the next subsection, Maxwell's equations can be expressed in the language of covariant classical field equations, which are developed in Chapter 9 of this book. (Covariant means that the equations are invariant under a Lorentz (or SR) transformation, which is described in Chapter 2 and Appendix A.) Moreover, the other incarnation of light, the photon, can be considered a zero-mass particle that moves at the speed of light. Then, from the general equations in SR for a particle of mass we simply take the limit $m \to 0$ (or alternatively $v \to c$) to obtain the equation for a photon. (For clarification, m is the mass of the particle, v is its speed, and c is the speed of light.) See Chapter 5. The photon picture is especially important for describing relativistic scattering problems (Chapter 7) and the relativistic Doppler shift (Chapter 8).

Another way of expressing these matters is to observe that SR is a classical theory of point-like particles with masses, which theory can also accurately describe the classical wave motion of the Maxwell field. However, by taking the zero-mass limit of a particle in SR we can obtain the photon formulation, which is the other representation of light. It should be mentioned too that there are other particles with zero mass—namely, the gluon and the graviton (if it exists). Moreover, years ago it was thought that neutrinos had zero mass before modern experiments established that the three varieties of neutrinos have very small but nonzero masses.

1.3.2. The Nineteenth-Century Ether (Aether) Theory of Light. Before the nineteenth century, physicists assumed the *Galilean theory of relativity*, which postulated that the laws of physics were the same in any *inertial frame of reference*, where an inertial frame is one that is moving uniformly (i.e., with no acceleration) with respect to another frame. Examples of two inertial frames are shown in Fig. A.1. The transformation law between inertial frames is given by the so-called *Galilean transformation*, which is discussed in Chapter 2 and Appendix A. Thus, Galilean relativity stated that Newton's laws of mechanics are invariant under a Galilean transformation. However, after the development of E&M in the nineteenth century, it was clear that Maxwell's equations were *not* invariant under a Galilean transformation.

For a clear demonstration of the differences in classical physics between mechanics and E&M, we recommend the award-winning physics website about SR, Einstein Light: `http://www.phys.unsw.`

---

*In 1905 Albert Einstein did the theoretical work on the photoelectric effect, for which he received the Nobel Prize in physics in 1921. However, Einstein never received a Nobel Prize for his very important, pioneering work on SR and GR.

`edu.au/einsteinlight`. Einstein Light has received a rave review in *Science* magazine and an award from *Scientific American*. It has several film clips (about a minute each), which use demonstrations and animations to give overviews on Galilean relativity, mechanics, electromagnetism, and the apparent inconsistency between Galilean relativity and electromagnetism.

Also, it was clear that Maxwell's equations predict a particular constant of propagation—namely, the speed of light. This constant appears as an apparent parameter of the theory, which can be seen most dramatically in the so-called classical "wave equations." Moreover, Galilean relativity predicted that the speed would change from reference frame to reference frame.

Thus, nineteenth-century physicists adopted a novel point of view—namely (i) classical relativity applied only to mechanics, not to E&M and (ii) Maxwell's equations were valid only in a special reference frame, the preferred frame moving with the mechanical, luminiferous ether. The ether was thought to be a massless, frictionless substance, pervading all space, through which the electromagnetic waves propagated. In those days, if you were told that light was a wave, the natural question was "Wave with respect to what?", and the natural answer was "The ether!". Also, it has been reported that Maxwell himself believed in the existence of the ether. It was said too that he eventually abandoned this idea, but unfortunately his early position strengthened the belief in the ether theory (just as Newton had doomed the corpuscular theory of light).

The next development took place toward the end of the nineteenth century. In 1887, Albert Michelson and Edward Morley reported their results determining the velocity of the Earth with respect to the ether. They had used an interferometer to measure velocities parallel and perpendicular to an assumed direction of the velocity of their apparatus with respect to the ether, and they had performed their experiment six months apart, so that—six months later—the effective relative ether velocity would be reversed in direction. The result was negative to their great surprise (and sadness because they really believed in the ether theory). *Therefore, there was no ether, and no preferred frame of reference!* Something was clearly wrong with Galilean relativity. Also, it was very clear that either the equations of Newton's mechanics or Maxwell's equations for E&M had to be modified.

Enter Einstein and SR in 1905. Einstein found that the equations of E&M were quite beautiful. Thus, what strongly motivated Einstein in developing SR was to save the elegant equations of Maxwell, even if it meant throwing Newton "under the bus".[†] Thus, it turned out that SR allowed the comparatively recent equations of Maxwell to remain intact, while Newton's Second Law (dating back more than two hundred years) had to be modified. In Einstein's SR, it is assumed that the laws of physics (all laws, not just mechanics) are the same in *any inertial frame* of reference. Also, Einstein made the revolutionary assumption that the speed of light was the same in any inertial frame, which has the consequence that *no speed can exceed the speed of light*. Thus, Einstein's SR uniquely solved all of the problems that arose during the nineteenth century due to the ether controversy.

Science historians appear to disagree as to what extent Einstein knew about the Michelson-Morley result. Some seem to imply that Einstein was only dimly aware (if at all) of this experiment.

[†]Einstein's original SR paper, entitled "On the Electrodynamics of Moving Bodies," can be found in "The Principle of Relativity: A Collection of Original Memoirs on the Special and General Theories of Relativity", trans. W. Perrett and G. B. Jeffry (Dover Publications), pg. 35: and was published in 1905 in the German journal *Annalen der Physik*. Notice that the title emphasizes electromagnetism rather than mechanics. Also, as shown in Chapter 5 of this book, in SR the famous equation $E = mc^2$ is derived, a result published in a separate paper in 1905. However, because this equation was published in a different paper, some people erroneously believe that there are two separate theories involved.

Others say that he knew about the experiment, which seems more likely since it was one of the main developments of the late nineteenth century.

Due to its revolutionary nature and the precise predictions it can make, SR was subjected to a multitude of tests during the twentieth century. Some of the most famous of these tests include measurements of the lifetimes of muons proving time dilation, the detailed predictions of accurate electron energy levels that depend on spin effects in hydrogen and other atoms (à la the Dirac equation, discussed in Chapter 12), the establishment of reliable beam dynamics in particle accelerators, tests proving the relativistic Doppler shift, and many experiments verifying the conversion of mass to energy, and vice versa. The upshot of all these experiments and reliable predictions is that SR is one of the best-tested scientific theories of all time. Next, in the coming chapters, we will delve into the details of SR.

Chapter 2

Inertial Frames; The Galilean And Lorentz Equations; Special Relativity

2.1. Inertial Frames

A coordinate system frame of reference can move uniformly with respect to another coordinate system. This movement can be a displacement, or a rotation, of one frame with respect to another. However, the motion with which we will be exclusively concerned is illustrated in Fig. A.1 in Appendix A; namely, one frame moves with a constant velocity $\vec{u}$ with respect to another frame and we define:

> An **inertial frame** is a system of coordinates that is at rest or moving with a *constant velocity*, $\vec{u}$, with respect to another coordinate system.

Note that *accelerated* relative motion of frames is *not* allowed.

We are interested in the transformation that maps the coordinates of one inertial frame given by (x, y, z, t) into those of a related inertial frame given by (x', y', z', t').

2.2. The Galilean Transformation

The classical transformation, in pre-relativity physics, is obtained in Appendix A.1 in Eqs. (A.1):

$$x' = x - ut \tag{2.1a}$$

$$y' = y \tag{2.1b}$$

$$z' = z \tag{2.1c}$$

$$t' = t, \tag{2.1d}$$

in which we assime that u is along the x and x' axes. (As mentioned in the Appendix, there is no loss of generality in this assumption). Also, it is assumed that at $t = t' = 0$, $\vec{r} = \vec{r'} = 0$ (with the origins of the two frames coinciding at the initial times). An important feature of the Galilean transformation is that $t = t'$, which implies that there is absolute time, which was widely accepted by Galileo, Newton, and many others. Also, we note that the coordinates y and z do not change, and that the inverse transformation is given by Eqs. (A.2) in Appendix A.

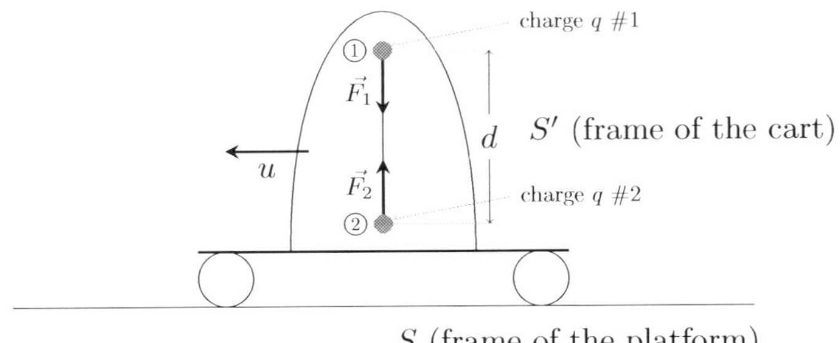

FIGURE 2.1. The two equal charges (q each) are separated by a distance d on a railroad cart (moving along the track to the left with velocity u). The magnetic forces F_1 and F_2 are observed only in Frame S.

We can also calculate velocities, using:

$$v_x = \frac{\mathrm{d}x}{\mathrm{d}t} \text{ and } v'_x = \frac{\mathrm{d}x'}{\mathrm{d}t'} \tag{2.2a}$$

$$v_y = \frac{\mathrm{d}y}{\mathrm{d}t} \text{ and } v'_y = \frac{\mathrm{d}y'}{\mathrm{d}t'} \tag{2.2b}$$

etc.

Then, from Eqs. (2.1) and (2.2), we obtain:

$$v'_x = v_x - u \tag{2.3a}$$

$$v'_y = v_y \tag{2.3b}$$

$$v'_z = v_z. \tag{2.3c}$$

Eqs. (2.3) can be inverted, giving

$$v_x = v'_x + u \tag{2.4a}$$

$$v_y = v'_y \tag{2.4b}$$

$$v_z = v'_z, \tag{2.4c}$$

a result that also follows from Eqs. (A.2). Note that in Galilean relativity the velocities are *additive*.

For now, we will let u denote a frame velocity and $v = \mathrm{d}x/\mathrm{d}t$, the velocity of a particle. Later, we will relax this distinction (in order to allow for the acceleration of particles).

2.3. Galilean Relativity

Galilean Relativity states that:

The laws of physics are the same in any inertial frame.

a. In *Newtonian mechanics*, this principal is obviously true. From Eqs. (2.2) and (2.3) we find that if

$$\vec{a} = \frac{\mathrm{d}^2\vec{r}}{\mathrm{d}t^2} = 0, \tag{2.5a}$$

this implies that

$$\vec{a}\,' = \frac{\mathrm{d}^2 \vec{r}\,'}{\mathrm{d}t'^2} = 0, \tag{2.5b}$$

which follows since $\vec{u}$ is a constant. Thus, if a particle feels no force in one inertial frame, it will feel no force in any other inertial frame. However, in a frame accelerating with respect to another, there would be a difference in the laws of mechanics.

b. *Galilean relativity applies to mechanics, not to electricity and magnetism.* See Fig. 2.1, in which a railroad cart is moving along a track with a velocity u. It contains two equal charges, each with value q, one at the top of the cart and the other at the bottom, separated by a distance d. An observer on the cart (frame S') and an observer at rest on the platform (frame S) each sees an electrostatic repulsion given by q^2/d (in cgs units). The observer on the cart sees *no* other force. However, the observer on the platform also sees a magnetic force of attraction between the two charges, which can be seen as follows. From the Biot-Savart Law, or simply from the right-hand-thumb rule for moving charges, we see that, in S, there is a magnetic field $\vec{B}_1$ at ① due to the moving charge at ② and this field is in a direction into the paper. Similarly, at ②, there is a magnetic field $\vec{B}_2$, in a direction out of the paper, from the moving charge at ①. We also note that $|\vec{B}_2| = |\vec{B}_1|$. Then, from the Lorentz force law we see that at ① there is a magnetic force proportional to $\vec{u} \times \vec{B}_1$, which points downwards. Similarly, at ②, there is a magnetic force, proportional to $\vec{u} \times \vec{B}_2$, pointing upwards. Thus, the laws of electricity and magnetism are different in the frames S and S', which is a violation of Galilean relativity. See the discussion on the web site "http://www.phys.unsw.edu.au/einsteinlight" of this behavior.

Thus, if you believed and insisted on the relativity principle, either mechanics or E&M had to be modified and/or discarded.

Physicists in the nineteenth century were quite willing to apply the Galilean relativity principal to mechanics, but not to electricity and magnetism. In particular, it was thought that electromagnetic waves propagated through a semi-mechanical medium, called the *ether*, which implied the existance of a *preferred frame of reference*.

In the late nineteenth century, Newton's Laws of Mechanics, which were about two hundred years old, were verified to be included in classical (Galilean) Relativity while the more recent Maxwell's equations were excluded from classical relativity. Thus, in the nineteenth century E&M became kind of a "second-class citizen" compared to mechanics. In contrast, Einstein believed that the equations of electricity and magnetism were very beautiful, and he searched for a theory that would preserve these laws, even at the expense of the violation of Newton's Laws. This Special Relativity (SR) theory was presented by Einstein[3] in 1905, and the mathematical basis for SR was the Lorentz transformation, which replaced the Galilean transformation.

2.4. Special Relativity and the Lorentz Equations

SR makes the following assumptions:

i) The laws of physics are the same in any inertial frame (not just mechanics, but electricity, magnetism and any other fundamental force of nature).

ii) In a vacuum, the speed of light, c, is constant in any inertial frame, which has the consequence that no speed can exceed the speed of light.

Using Assumption (ii) in Appendix A, we derive the Lorentz transformation, which is given by

$$x' = \gamma(x - ut) \tag{2.6a}$$

$$y' = y \tag{2.6b}$$

$$z' = z \tag{2.6c}$$

$$t' = \gamma\left(t - \frac{u}{c^2}x\right), \tag{2.6d}$$

where

$$\gamma(u) = \left[1 - u^2/c^2\right]^{-1/2} = \gamma(-u). \tag{2.7}$$

In Appendix A, we also show that a by-product of the derivation is that the magnitude of the frame velocity, $|u|$, can not exceed the speed of light. We note too that in SR, $t' \neq t$, so that there is no absolute time. Also, the inverse transformation to Eqs. (2.6) is given in Eqs. (A.14) of Appendix A.

Next, we take differentials of Eq. (2.6), obtaining

$$dx' = \gamma(dx - u\,dt) \tag{2.8a}$$

$$dy' = dy \tag{2.8b}$$

$$dz' = dz \tag{2.8c}$$

$$dt' = \gamma\left(dt - \frac{u}{c^2}dx\right). \tag{2.8d}$$

Then, from Eqs. (2.2) we find for particle velocities that

$$v'_x = \frac{dx'}{dt'} = \frac{dx - u\,dt}{dt - \frac{u}{c^2}dx} = \frac{v_x - u}{1 - \frac{u}{c^2}v_x}, \tag{2.9a}$$

where $v_x \equiv dx/dt$, $v'_x \equiv dx'/dt'$, etc, and

$$v'_y = \frac{v_y}{\gamma\left(1 - \frac{u}{c^2}v_x\right)} \tag{2.9b}$$

$$v'_z = \frac{v_z}{\gamma\left(1 - \frac{u}{c^2}v_x\right)}. \tag{2.9c}$$

Notice that, in contrast to Eq. (2.4), the perpendicular components of $\vec{v}\,'$ are not trivial. Also, it is clear that in SR the velocities are no longer additive.

We now give a particular example showing that you can not transform to a frame where the magnitude of a velocity exceeds the speed of light. The inverse transformation to Eq. (2.9a) is obtained by letting $u \to -u$, so that

$$v_x = \frac{v'_x + u}{1 + \frac{u}{c^2}v'_x}, \tag{2.10}$$

and we take the special case

$$0 < v'_x \leq c \tag{2.11a}$$

$$0 < u < c, \tag{2.11b}$$

the last of which follows from the derivation of the Lorentz transformation in Appendix A. From Eq. (2.10), we find that

$$v_x = c\left\{\frac{\frac{v'_x}{c} + \frac{u}{c}}{1 + \frac{u}{c^2}v'_x}\right\},$$

which can be re-expressed as

$$v_x = c \left\{ \frac{1 + \frac{u}{c^2} v'_x - 1 - \frac{u}{c^2} v'_x + \frac{v'_x}{c} + \frac{u}{c}}{\left(1 + \frac{u}{c^2} v'_x\right)} \right\} \tag{2.12}$$

or

$$v_x = c \left[1 - \frac{(1 - \frac{v'_x}{c})(1 - \frac{u}{c})}{(1 + \frac{u}{c^2} v'_x)} \right].$$

However, from Eqs. (2.11)

$$0 \le 1 - \frac{v'_x}{c} < 1$$

$$0 < 1 - \frac{u}{c} < 1$$

$$1 + \frac{u}{c^2} v'_x = 1 + \left(\frac{u}{c}\right)\left(\frac{v'_x}{c}\right) > 1,$$

so that the bracketed term in Eq. (2.12) is less than unity, and

$$v_x \le c. \tag{2.13}$$

Thus, if we start out with a speed less than c we can not transform to a frame in which the speed exceeds c. The restrictions in Eqs. (2.11) can be removed, so that if $|v'_x| \le c$, then in general $|v_x| \le c$. Also, other examples are given in the Chapter exercises.

It is useful to summarize the main results of Appendix A and this chapter regarding the assumption that the speed of light is the same in any inertial frame. First, in Appendix A, it was shown that the assumption led to the famous Lorentz transformation, with the added condition that the coefficients of the Lorentz transformation are real and finite if and only if the speed of the moving frame of reference is less than the speed of light. Then, in this chapter we have obtained the transformation equations for the Lorentz particle velocities (Eqs. (2.9)), after which we demonstrate that the limitation on the frame speed coupled with the velocity transformation equations leads, for a particular case, to the condition that the speed of any particle can not exceed the speed of light. (Eq. (2.13)) We then assert that this is a general behavior. Thus, the speed of any object (frame or particle) can not exceed the speed of light.

When the paper on SR was published by Einstien in 1905, it shook the world of physics to its very core. Several features of the new theory overthrew various aspects of the older nineteenth-century physics. Other features were non-intuitive, leading to effects that were counter to previous models of physical behavior.

We now summarize why SR was considered revolutionary in 1905.

(1) *The idea that* all *laws of physics are included in the relativity principle (not just mechanics).* This also led to the abolishment of the *ether*, which—in turn—meant that there was no preferred frame of reference.

(2) *The constancy of the magnitude of the velocity of light, c, in any inertial frame, which is not at all intuitive.* As mentioned, this implies that no particle or frame speed can exceed the speed of light.

(3) *The idea of mixing space and time, so that what is space in one frame partially becomes time in another, and vice versa.* This implies that neither absolute space nor absolute time exists. We now speak of the existence of the *SpaceTime Continuum*.

(4) $E = mc^2$, *a relation that we will prove later**. This famous equation extended energy conservation to include mass. Previously, the science of chemistry was built on the foundation that mass and energy are separately conserved. Now, all forms of energy, including mass, are included under the same law. The equation $E = mc^2$ has some important consequences.

    (a) First, a small amount of mass can be converted into a large amount of energy (e.g. nuclear bombs or nuclear reactors). At the time of the first atomic (i.e. nuclear) bomb test explosions in New Mexico, the physicists involved were very surprised, even shocked, at the amount of energy released. Some of these physicists became very opposed to using the new weapon in World War II, especially after Nazi Germany surrendered. Then, after the subsequent bombings of the two cities in Japan, many physicists had a collective guilt about unleashing this new form of energy. Since World War II, no nuclear bombs have been used in warfare.

    (b) Second, small masses, moving at high speeds (so that they possess a lot of energy) can collide and annhilate, producing a particle (or particles) with a very large mass. For example, in the e^+ and e^- annhilation in the CERN accelerator, the very massive Z° boson was created. The Z° has a mass $\approx 92,000$ times larger than the sum of the e^+ and e^- masses. Truly, this reaction converts energy into mass with a vengence. See subsection 7.3.3.

---

*In 1905 the $E = mc^2$ paper was published separately by Einstein from his main paper on SR. Thus, there has been a tendency to regard these two papers as separate theories. In truth, $E = mc^2$ is—as we shall see in Chapter 5—part of SR (and derived from SR).

(1) Assume the Lorentz transformation, Eqs. (2.6), relating the coordinates and time in the inertial frame of reference S' to those in the inertial frame S. Then, find the inverse transformation; i.e., the transformation relating the coordinates and time in S' to those in S (Eqs. (A.14) in Appendix A). Note that the Lorentz equations are *linear equations*, relating x' and t' to x and t, so it is fairly easy to find the "inverse transformation", expressing x and t in terms of x' and t'. You can very easily solve this problem just by reversing the sign of u, as we do in Appendix A, but we are asking here for you to solve the linear equations.

(2) Assume that an inertial frame S' is moving along the positive x-axis with velocity $u = (4/5)c$ relative to a fixed inertial frame S and that the origins of their coordinate systems coincide initially. (Note: $c = 3.0 \times 10^{10}$ cm/sec is the velocity of light in a vacuum.) Find the following:

    (a) If the coordinates and time of an event in S are given by $x = 0.3$ cm, $y = 0.5$ cm, $z = 0.8$ cm, $t = (1/3) \times 10^{-10}$ sec, find the coordinates and time in the S' frame.

    (b) If the coordinates and time of an event in S' are given by $x' = 0.2$ cm, $y' = 0.8$ cm, $z' = 0.4$ cm, $t' = (1/9) \times 10^{-10}$ sec, find the coordinates and time in the S frame.

Note: in parts a and b you are asked to use the Lorentz transformation and the inverse Lorentz transformation, respectively. Notice the units of the time variables and the magnitude of c, which *suggests that a formulation in terms of ct and ct' would be useful*. Also, it is useful to express *all velocities in terms of ratios with respect to c*.

(3) This problem demonstrates the impossibility of achieving a velocity greater than light by transforming to a different inertial frame of reference.

    Assume that an inertial frame S' is moving along the positive x-axis with velocity u relative to a fixed inertial frame S and that the origins of their coordinate systems coincide initially. Note: use the *inverse transformation of velocities* in both parts of the problem. It is helpful to express all velocities in terms of ratios with respect to c.

    (a) Assume that $u = (3/4)c$ and that Superman in S' is running at half the speed of light in the x' direction. (Of course, Superman could run even faster, but he cannot *exceed* the speed of light!) What is the speed of Superman as observed in the frame S?

    (b) An observer in S' turns on a flashlight, emitting light (which travels at the speed of light in a vacuum, c). (Note: $v'_x = c$.) Show that in S the velocity of light emitted from the flashlight is the velocity of light, c.

Chapter **3**

Plots of ct' vs. x' and Three Important Applications of Special Relativity

3.1. ct' vs. x' plots

In Eqs. (A.16) of Appendix A we write the Lorentz equations in four-vector form (to be discussed shortly), relating ct and x to ct' and x'.

Next, we note the interpretation of the line in a ct vs. x plot, shown in Fig 3.1. The ct-line is the locus of points for which $x = 0$, and the x-line is locus of points for which $ct = 0$, or $t = 0$. Then, lines parallel to the ct-axis denote position, or events, in the $ct - x$ frame, which do not change with time (t), whereas lines parallel to the x-axis denote events occurring at the same time (t) in the $ct - x$ frame. (I.e., for the latter, all points along such a line are *simultaneous* in this frame. Simultaneity will be discussed later in this chapter.)

Moreover, in analogy with the $ct - x$ axes, the ct'-line is defined by the equation $x' = 0$ while the x'-line is defined by $ct' = 0$. This, using the two equations (A.16a) and (A.16d), we find that

$$x' = \gamma \left[x - \frac{u}{c}(ct) \right] = 0 \qquad \text{(Equation for the } ct'\text{-line)} \tag{3.1a}$$

and

$$ct' = \gamma \left[ct - \frac{u}{c}x \right] = 0 \qquad \text{(Equation for the } x'\text{-line)}, \tag{3.1b}$$

or

$$ct = \frac{u}{c}x \qquad \text{(Equation for the } x'\text{-line)} \tag{3.2a}$$

and

$$ct = \frac{c}{u}x \qquad \text{(Equation for the } ct'\text{-line)}. \tag{3.2b}$$

Next, in Fig. 3.2 we plot the ct' and x' lines on a $ct - x$ diagram. First, we see that the x' line has a slope of

$$0 < \phi = \tan^{-1}\left(\frac{u}{c}\right) < \frac{\pi}{2}, \tag{3.3}$$

while the ct' line has a slope of

$$\alpha = \tan^{-1}\left(\frac{c}{u}\right) = \cot^{-1}\left(\frac{u}{c}\right) = \frac{\pi}{2} - \lambda, \tag{3.4}$$

which implies that

$$\lambda = \phi. \tag{3.5}$$

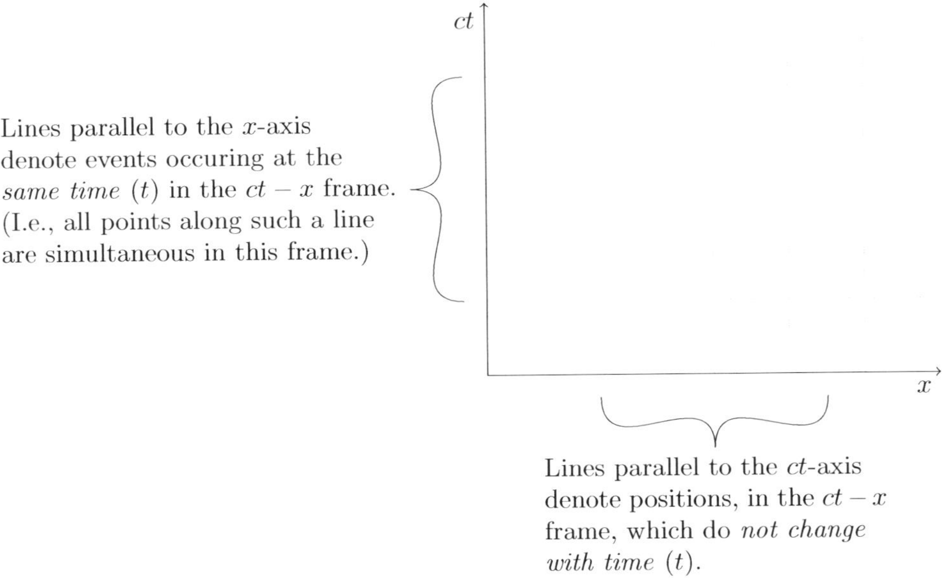

FIGURE 3.1. *ct* vs *x* plot

Then, see Fig. 3.3 which gives the interpretation of lines parallel to the x' and ct' axes on a *ct* vs. *x* plot. We see that lines parallel to the x'-axis denote events occurring at the same time (t') in the $ct' - x'$ frame. Also, lines parallel to the ct'-axis denote events in the $ct' - x'$ frame which do not change with the time t' and, thus, are simultaneous in this frame.

See Problem 1. for this chapter, which is the basis for Fig. 3.4. We see that ϕ' is a negative angle, with

$$|\phi'| = \phi. \tag{3.6}$$

Also, we have

$$\lambda' = \phi. \tag{3.7}$$

One feature that is evident from Figs. (3.2), (3.3), and (3.4) is that it is not possible to change the sign of t or x by a proper Lorentz transformation. As one book emphasizes *"Proper Lorentz transfomations do not interchange past and future"**

3.2. Simultaneity

We have previously discussed simultaniety in the $ct - x$ and $ct' - x'$ frames. (See also Fig. 3.5). If two events, which are points, in the frame, occur at the same frame time, they are said to be simultaneous.

However, in Special Relativity, two events that are simultaneous in one intertial frame are *not* simultaneous in other inertial frames.

---

*An improper Lorentz transformation might include the operations of time reversal, a spatial reflection, or a spatial inversion, all of which we exclude from the formalism in this book.

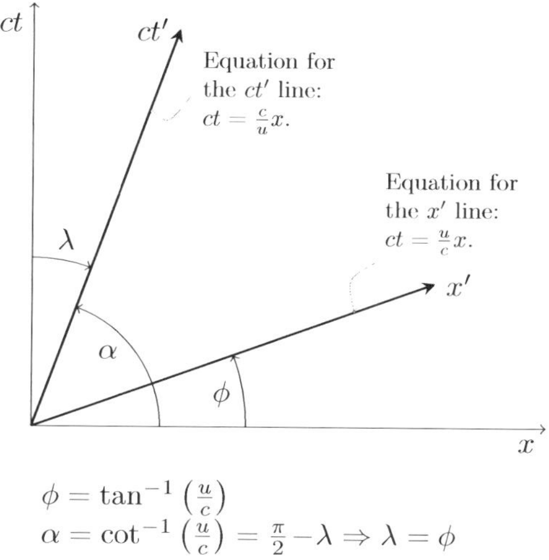

$$\phi = \tan^{-1}\left(\tfrac{u}{c}\right)$$
$$\alpha = \cot^{-1}\left(\tfrac{u}{c}\right) = \tfrac{\pi}{2} - \lambda \Rightarrow \lambda = \phi$$

FIGURE 3.2. Plot of the ct' and x' lines on a $ct - x$ diagram.

This behaviour can be seen from Eq. (2.6d), for which we have, for two events ar (x_1, t_1) and (x_2, t_2):

$$t_2' - t_1' = \gamma\left[(t_2 - t_1) - \frac{u}{c^2}(x_2 - x_1)\right]. \tag{3.8}$$

Now, if they are simultaneous in the unprimed frame, $t_1 = t_2$, but from Eq. (3.8) we see that

$$t_2' \neq t_1' \tag{3.9}$$

unless $x_1 = x_2$, which is a trivial case, occurring only if the two points (or events) are identical.

Moreover, we see from the Galilean transformation, Eq. (A.1d), that if $t_1 = t_2$ then $t_1' = t_2'$, so that two events simultaneous in one inertial frame are simultaneous in all inertial frames. Clearly, this difference in behavior between Galilean physics and SR arises because $t' = t$ in the former, but $t' \neq t$ in the latter.

3.3. Time Dilation

For the inverse transformation, Eq. (A.14d), we have

$$t = \frac{t' + \frac{u}{c^2}x'}{\left(1 - \frac{u^2}{c^2}\right)^{\frac{1}{2}}}, \tag{3.10}$$

for which we can write, for two events at points 1 and 2:

$$t_2 - t_1 = \frac{(t_2' - t_1') + \frac{u}{c^2}(x_2' - x_1')}{\left(1 - \frac{u^2}{c^2}\right)^{\frac{1}{2}}}. \tag{3.11}$$

Assume that there is a clock in S' that is *stationary*, so that $x_2' = x_1'$. Then, we have

$$t_2 - t_1 = \frac{t_2' - t_1'}{\left(1 - \frac{u^2}{c^2}\right)^{\frac{1}{2}}} = \frac{\tau_2' - \tau_1'}{\left(1 - \frac{u^2}{c^2}\right)^{\frac{1}{2}}}, \tag{3.12}$$

where τ is the time recorded on the stationary clock, which we refer to as the *proper time*. This is a statement of *time dilation*, which says that

> A clock at rest in S' appears to be running more slowly, as seen by an observer in S.

For example, if $\frac{u}{c} = \frac{\sqrt{3}}{2}, \gamma = (1 - \frac{u^2}{c^2})^{-\frac{1}{2}} = 2$ and if $t'_2 - t'_1 = 1$ hour, the proper time interval recorded on the clock at rest in S', and $t_2 - t_1 = 2$ hours, the time interval recorded by the stationary observer in S. Thus, if S' is a space ship, the man aboard will have aged 1 hr, while the man in S (say, on Earth), will have aged 2 hrs. If the two persons are twins, this gives the famous *Twin Paradox*, whereby one twin goes off in a spaceship and returns younger than his twin on Earth.

A naïve interpretation of this scenario would reason that, by relativity, each twin would experience himself at rest and the other twin in motion, and would thus see his counterpart aging less rapidly than himself. The two twin's situations appear to be completely reciprocal. However, the two situations are *not* really reciprocal. The twin on the space ship is doing something quite different from the Earth-bound twin. In order to turn around and return to Earth, the space-ship twin must undergo a deceleration and then an acceleration. (Also, there is an acceleration when he leaves Earth and a deceleration when he returns.) It used to be said that the Twin Paradox could only

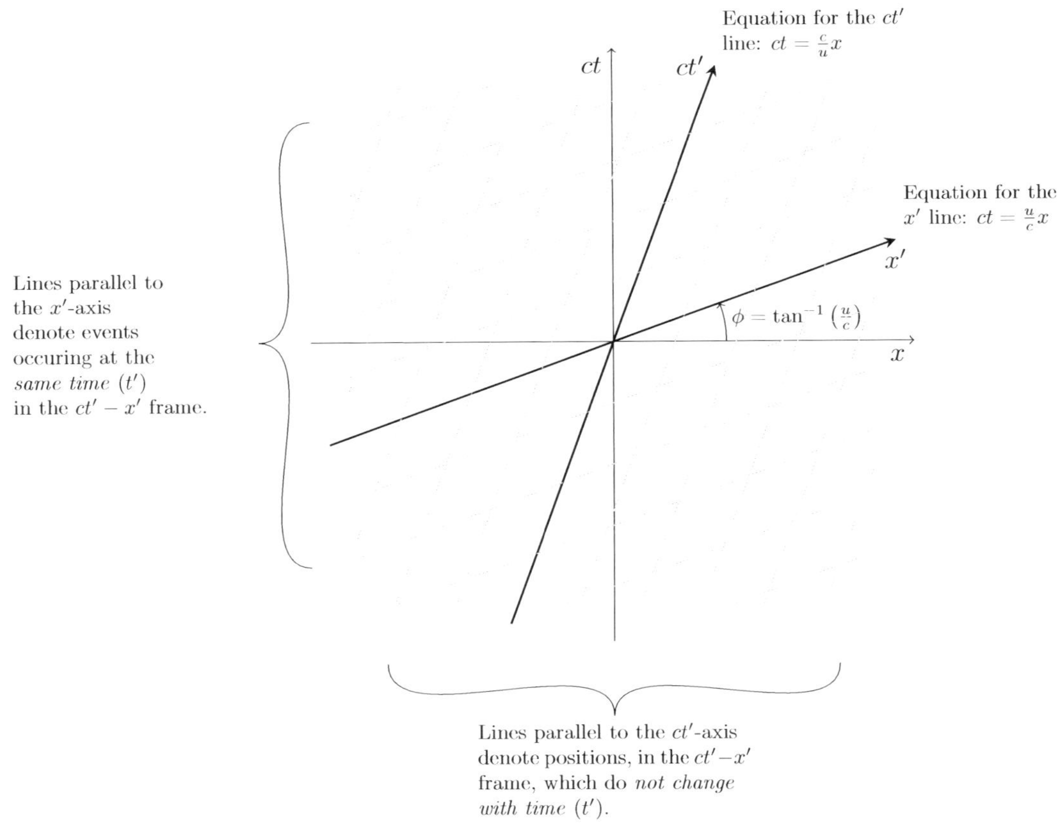

FIGURE 3.3. Lines parallel to the x' and ct' axes on a ct vs x plot.

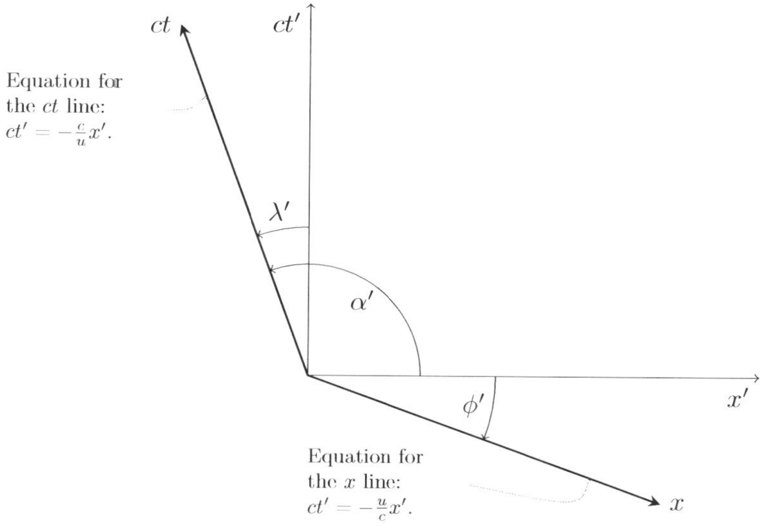

FIGURE 3.4. Plot of the ct and x lines on a ct' vs. x' diagram.

be resolved using General Relativity, which deals with accelerated motion. But this is an old wives' tale since, as we shall see, the paradox can be resolved completely within SR. We shall demonstrate the full proof of this behavior in the next subsection using a simple SpaceTime diagram.

The resolution of the Twin Paradox illustrates another important point about SR. Namely, you can travel into the future since obviously the twin on Earth represents the future which would have existed for the space-ship twin. However, as we shall demonstrate in the next chapter, in SR you can not travel into the past, which would violate causality and necessitate a speed greater than the velocity of light.

We can also see time dilation using the $ct - x$ diagram shown in Fig. 3.6. We have (dividing through by the constant c)

$$\Delta t = t_W - t_P = t_E - t_M,$$

from which Eq. (A.14d) becomes

$$\Delta t = \gamma \left[(t'_E - t'_M) + \frac{u}{c^2}(x'_E - x'_M) \right].$$

However, $x'_E = x'_M$, so that

$$\Delta t = \gamma(t'_E - t'_M) = \gamma \Delta t', \tag{3.13}$$

which is time dilation, Eq. (3.12).

Important applications of time dilation occur in particle physics and in cosmology.

3.3.1. Muon Decay. The muon is a lepton similar to the electron but heavier. Whereas the electron is completely stable, the muon is unstable and decays after a short time. Muon decay has been verified experimentally. There are actually two muon decays, namely, for the $+$ and $-$ charged muons:

$$\mu^- \to e^- + \bar{\nu}_e + \nu_\mu \tag{3.14a}$$

$$\mu^+ \to e^+ + \nu_e + \bar{\nu}_\mu \tag{3.14b}$$

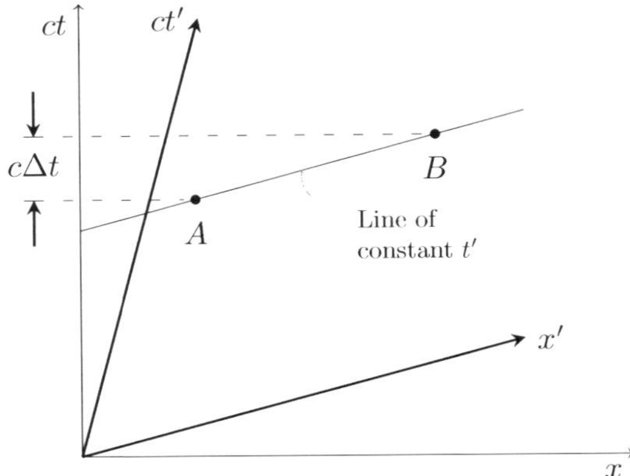

FIGURE 3.5. Illustration of simultaneity in the primed frame and lack of simultaneity in the un-primed frame. Events A and B are simultaneous in the (ct', x') frame because they occur at the same value of t'. They are *not* simultaneous in the (ct, x) frame, where A occurs before B.

where e^- is an electron, e^+ is a positron, ν_e and ν_μ are the electron and muon neutrinos, respectively, and the bars represent anti-neutrinos. Since μ^+ and μ^- are anti-particles of one another, they have identical decay times. Notice too that Eq. (3.14a) is converted to Eq. (3.14b) by changing each particle in the equation to its anti-particle (by formally using the charge-conjugation operator).

The muon decay time observed experimentally is given by Eq. (3.12), with

$$\Delta t = \frac{\tau_\circ}{\sqrt{1 - \beta^2}}; \quad \beta = \frac{v}{c}, \tag{3.15}$$

where $\tau_\circ$ is the *intrinsic* decay time of the muon. Also, the distance travelled before the muon decays is called the decay length and is given by

$$d = \Delta t v$$
$$= \frac{\tau_\circ}{\sqrt{1 - \beta^2}} \beta c. \tag{3.16}$$

It has been determined that

$$\tau_\circ = 2.2 \times 10^{-6} \text{sec}. \tag{3.17}$$

Then for most muon decays observed on Earth $\beta \approx .67$, which means that $d \approx 600$ meters.

Some muons are produced when cosmic rays (often very high energy protons) interact with the upper atmosphere. These muons travel at velocities given by $\beta \approx .997$, which gives $d \approx 8.5$km. Thus, many of these muons reach the surface of the Earth before they decay. In fact, this phenomenon was precisely how muons were discovered in 1937 in cosmic-ray research with cloud chambers (by Carl Anderson who also discovered the positron in 1931 in his cloud chamber).

Then, in a special muon acceleration experiment at CERN (the European collider) in the late 1970s, muons were accelerated to

$$\beta \sim .9994.$$

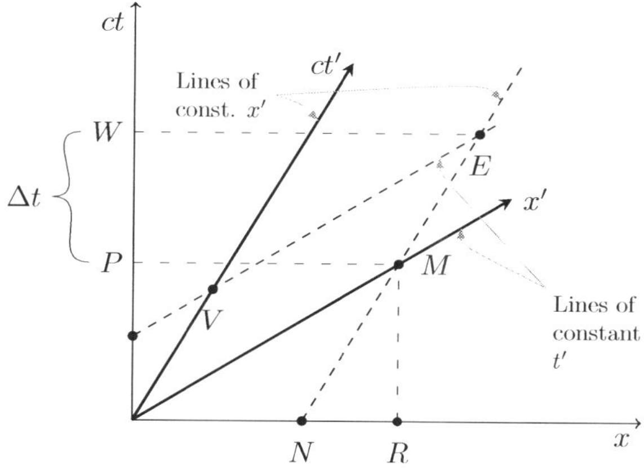

FIGURE 3.6. Time dilation as seen from a $ct - x$ diagram

The decay of these extremely high energy muons was observed to occur after

$$\gamma\tau_\circ \sim 64 \times 10^{-6}\,\text{sec}.$$

at a distance of

$$d \sim 19\,\text{km}.$$

Thus, time dilation has been very well tested experimentally.

3.3.2. Measurements in Cosmology on the Acceleration of the Universe[†]. In the measurements showing that the universe is accelerating, astronomers have to measure the brightness of Type Ia Supernovae, where it has been established that there is a relationship between intrinsic brightness and the rate that the Supernova's brightness changes over time (the Supernova's *light curve*). Once the intrinsic brightness has been determined, one can use the difference between the apparent brightness and intrinsic brightness to determine the distance to the galaxy in which the supernova resides.

To get an accurate measure of intrinsic brightness, astronomers also have to take into account time dilation effects due to the expansion of the Universe, i.e.: the light-curve includes two effects: (1) the Supernova's natural light curve and (2) a correction due to the expansion of the Universe. This implies that the measured lifetime of the supernova is actually longer than its actual lifetime in its rest frame.

3.4. The Differential of Proper Time

The differential of proper time, τ, is defined in Eq. (A.17) of Appendix A:

$$c^2\mathrm{d}\tau^2 = c^2\mathrm{d}t^2 - \mathrm{d}r^2, \tag{3.18}$$

where

$$\mathrm{d}r^2 = \mathrm{d}x^2 + \mathrm{d}y^2 + \mathrm{d}z^2. \tag{3.19}$$

[†]See Ref. 35, page 79.

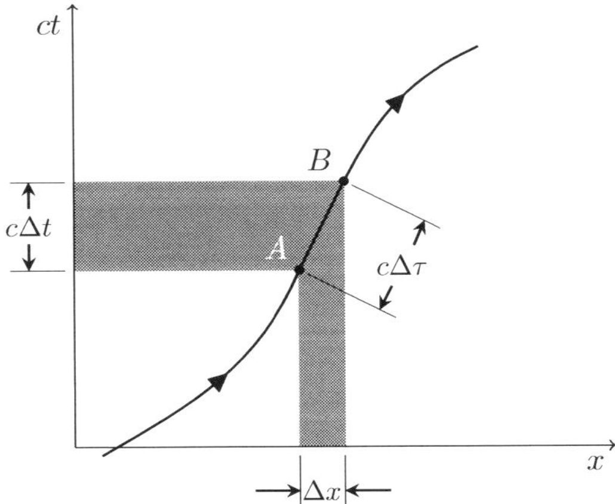

FIGURE 3.7. A one-dimensional world line in SpaceTime. The particle moves in the x-direction. The curve shown displays the increments $\Delta x, c\Delta t$, and $c\Delta\tau$, between points A and B. Appearances to the contrary, Δt is longer than $\Delta\tau$, which is time dilation. Note that this geometry is not Euclidean. Also, recall that $\Delta\tau$ refers to a clock moving with the particle.

In Appendix A, we showed that $d\tau^2$ is invariant under a Lorentz transformation. Recall too the definition of a proper time given earlier; namely the time recorded on a clock at rest in a particular frame. For such a clock, from Eq. (3.18), $dr^2 = 0$ and $d\tau = dt$. We can also rewrite Eq. (3.18) as

$$
\begin{aligned}
d\tau^2 &= dt^2 - \frac{d\vec{r}\cdot d\vec{r}}{c^2} \\
&= dt^2 \left[1 - \frac{1}{c^2}\frac{d\vec{r}}{dt}\cdot\frac{d\vec{r}}{dt} \right] \\
&= dt^2 \left(1 - \frac{v^2}{c^2} \right),
\end{aligned}
\tag{3.20}
$$

where

$$
\vec{v} = \frac{d\vec{r}}{dt}; \quad v = |\vec{v}| = (\vec{v}\cdot\vec{v})^{\frac{1}{2}},
\tag{3.21}
$$

or

$$
d\tau = dt \left(1 - \frac{v^2}{c^2} \right)^{\frac{1}{2}}.
\tag{3.22}
$$

Eq. (3.22) is also time dilation if we compare it with Eq. (3.12). However, in Eq. (3.22) v is more akin to a *particle velocity* (in a particular frame, as we see from Eqs. (2.2)), whereas in Eq. (3.12), u is a *frame* velocity. Thus, for an object, or a particle, following a *world line* in *SpaceTime*, as in Fig. 3.7, we imagine a series of *instantaneous inertial frames*, each with velocity $u = v$. This also has the advantage of allowing us to treat the *acceleration* of particles in SR.

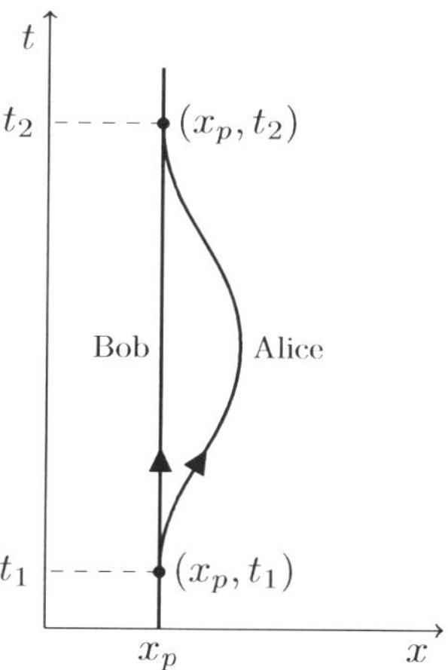

FIGURE 3.8. World lines of two twins, illustrating the twin paradox. Bob is the stay-at-home twin, while Alice goes off in a spaceship following the world line shown.

We can calculate the total proper time between two SpaceTime points, as in Fig. 3.7. From Eq. (3.22), we have

$$\Delta\tau = \tau_{AB} = \int_A^B d\tau = \int_A^B \sqrt{1 - (v^2/c^2)}\, dt, \tag{3.23}$$

and we see that in general

$$\Delta\tau < \Delta t \tag{3.24}$$

for finite or infinitesimal increments, which once again is time dilation.

In Fig. (3.8) we return once again to the twin paradox. Clearly, from Eq. (3.24), the proper time recorded on Alice's wristwatch is less than that recorded by Bob. Hence, in general, on reuniting at t_2, Bob will have aged more than Alice. We have then resolved the twin paradox for the general case. The travelling twin will always age less than the stay-at-home twin, an effect due to *time dilation* in SR. (Thus, there is no need to invoke General Relativity.)

3.5. The Lorentz Contraction

First, we shall give a general statement of the effect and then offer two general proofs.

Assume that a ruler in the x-direction is fixed in a moving frame, S', moving with a velocity u, in the x-direction, with respect to another frame S. If the ruler has a length in S' of l', in S it will have a length l, given by

$$l = \sqrt{\left(1 - \frac{u^2}{c^2}\right)}\, l'. \tag{3.25}$$

This is called a *Lorentz Contraction*.

Please note that ruler lengths perpendicular to the direction of motion do not change. This behavior is, of course, a consequence of the Lorentz equations (2.6), for which the perpendicular coordinates do not change.

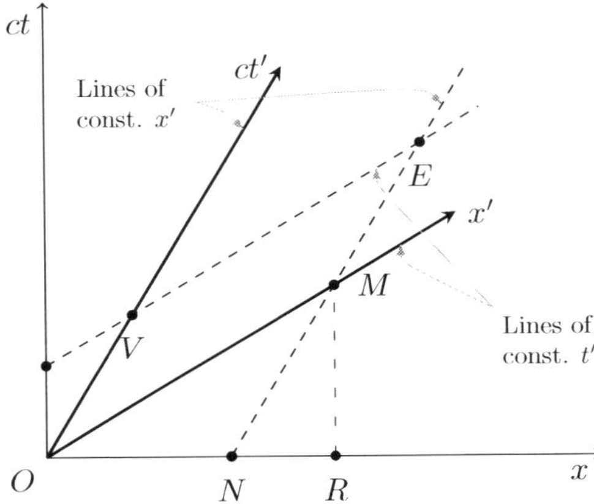

FIGURE 3.9. We will illustrate the Lorentz contraction for observers in S and S'.

We shall give two proofs of Eq. (3.25), first for an observer in S, measuring a ruler l' in S', and then for an observer in S', measuring a ruler l in S. For the first case, see Fig. 3.9, for which we see that the length $\overline{OM} = x'_M$ is constant with time (t') in the $ct' - x'$ frame. However, from Eq. (2.6a) we have

$$x'_M = x'_N = \gamma(x_N - ut_N).$$

But $t_N = 0$, so that

$$x'_M = \gamma x_N,$$

or

$$x_N = \sqrt{1 - \frac{u^2}{c^2}}\, x'_M = \sqrt{1 - \frac{u^2}{c^2}}\, x'_N,$$

which is the Lorentz contraction meansured in the unprimed frame.

Similarly, in Fig. 3.9, the length $\overline{OR} = x_R$ is constant with time (t) in the $ct - x$ frame. However, from Eq. (A.14a) we have

$$x_R = x_M = \gamma(x'_M + ut'_M).$$

But $t'_M = 0$, so that

$$x_R = x_M = \gamma x'_M,$$

or

$$x'_M = \sqrt{1 - \frac{u^2}{c^2}}\, x_M,$$

which is the Lorentz contraction measured in the primed frame.

From the analysis of Fig. 3.9, notice that $x'_M < x_R$, despite the appearence of the figure. This effect is again due to the non-Euclidean geometry of SpaceTime.

In Appendix B we give another derivation of the Lorentz contraction.

3.6. Problems

(1) **Plot the x and ct lines in the $ct' - x'$ coordinate system.**

In the text, we used the Lorentz Equations to plot the x' and ct' lines in the $ct - x$ coordinate system. In this coordinate system, the ct-axis is the vertical axis and the x-axis is the horizontal axis. We now want to use the Inverse Lorentz Equations to plot the x and ct lines in the $ct' - x'$ coordinate system, where the ct'-axis is the vertical axis, and the x'-axis is the horizontal axis.

Procedures:

(a) Find the equation for the x-line by setting $t = 0$ in the Inverse Lorentz Equations. This gives an equation of the form $ct' = f(x')$, where f is a linear function.

(b) Find the equation for the ct-line by setting $x = 0$ in the Inverse Lorentz Equations. This gives an equation of the form $ct' = g(x')$, where g is a linear function.

(c) Perform some geometry and give a qualitative plot of the x and ct lines in the $ct' - x'$ coordinate system. You should be able to reproduce Fig. 3.4, with

$$\lambda' = -\phi' = \phi$$
$$\alpha' = \cot^{-1}(-u/c).$$

(2) A twin is in a spaceship, moving with a constant velocity, $u = \frac{3}{5}c$, with respect to his Earth-based twin brother. The spaceship eventually turns around and moves with the same speed, eventually returning to Earth. We will assume that the spaceship makes instantaneous decelerations and accelerations. When the twins meet after the long separation, the twin who had travelled has aged 40 years. How much has his twin brother aged?

(3) One twin stays on Earth and the other goes off in a spaceship. At some point, the space-faring twin gets homesick and decides to return home. Thus, the trip has two legs that are shown on the accompanying diagram, Figure 3.10. Assume instantaneous accelerations and decelerations at A, B, and Q. On the *first leg* of the trip, the spaceship travels at a speed of $u = \frac{4}{5}c$ and the *Earth-bound twin* ages 30 years. For the *second leg* of the trip, the spaceshiup travels at a speed of $u = \frac{24}{25}c$ (since the twin is really anxious to return home) and the Earth-bound twin ages 25 years. Thus, the Earth-bound twin will have aged 55 years. How much will the space-faring twin have aged when the two twins are reunited?

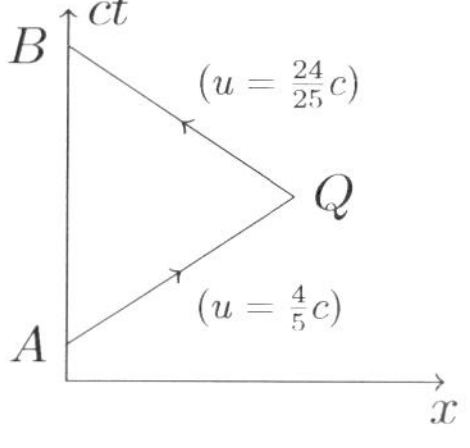

FIGURE 3.10.

25

(4) A train is moving with a velocity $u = \frac{3}{5}c$, with respect to an observer on the platform. Assume that a ruler affixed to the train measures 10cm. What is the length of the ruler as measured by the observer on the platform?

Chapter **4**

Four-Vectors, Light Cones, Causality, Time Travel, and Relativistic Vector Calculus

4.1. Four Vectors

As mentioned in Appendix A (see Eq. (A.15)), we can write space and time coordinates in four vector form as follows:

$$X = X_\mu = (X_1, X_2, X_3, X_t) = (x_1, x_2, x_3, x_t) = (x, y, z, ct) = (\vec{r}, ct), \tag{4.1a}$$

which can be rewritten as

$$X = (\vec{r}, X_t). \tag{4.1b}$$

Notice that the time-like part $(X_t = ct)$ of the four vector has the same units as the space-like part $(\vec{r} = x, y, z)$. Also, the space-like part has three terms. We will use the notation X_μ sometimes to refer to the total four vector X (as above), or to refer to a particular component: $\mu = 1, 2, 3, t =$ x, y, z, t. Some books use 0 or 4 for the time component; we prefer simply t.

As mentioned in Appendix A, the components of the four vector obey the Lorentz transformation; namely,

$$x'_1 = \gamma \left(x_1 - \frac{u}{c} X_t \right) \tag{4.2a}$$

$$x'_2 = x_2 \tag{4.2b}$$

$$x'_3 = x_3 \tag{4.2c}$$

$$X'_t = \gamma \left(X_t - \frac{u}{c} x_1 \right), \tag{4.2d}$$

with

$$\gamma = \left(1 - \frac{u^2}{c^2} \right)^{-\frac{1}{2}}. \tag{4.3}$$

As we shall see, there are many other four vectors that can be constructed and we write each of these as

$$A = A_\mu = (A_x, A_y, A_z; A_t) = (\vec{A}, A_t), \tag{4.4}$$

so that there is a three-component spatial part $(\vec{A})$ and a one-component time-like part (A_t). Thus, $\vec{A}$ plays the role of $\vec{r}$ and A_t, that of ct for the SpaceTime coordinates in (4.1a). For the four vector

A and its components, sometimes it will be convenient to adopt a lower-case letter and at other times a capital letter. The general four vector in Eq. (4.4) then satisfies the Lorentz equations:

$$A'_1 = \gamma \left(A_1 - \frac{u}{c} A_t \right) \tag{4.5a}$$

$$A'_2 = A_2 \tag{4.5b}$$

$$A'_3 = A_3 \tag{4.5c}$$

$$A'_t = \gamma \left(A_t - \frac{u}{c} A_1 \right). \tag{4.5d}$$

Recall that, as before, the primed system S' moves with a velocity u in the x-direction, with respect to the stationary frame S. Also, notice the symmetry between Eqs. (4.5a) and (4.5d).

4.2. Operations With Four Vectors: the Metric for SR

Suppose we combine ($\pm$) one four vector A with another B to form a third four vector C:

$$C = A \pm B. \tag{4.6}$$

Then, in Eq. (4.6), for the components, we have

$$C_\mu = A_\mu \pm B_\mu \tag{4.7}$$

Moreover, if we multiply a four vector A by a constant k

$$B = kA, \tag{4.8}$$

the components satisfy

$$B_\mu = k A_\mu. \tag{4.9}$$

We can also construct a four-dimensional dot-product (or inner product) as follows:

$$A \cdot B = \vec{A} \cdot \vec{B} - A_t B_t \tag{4.10}$$

where

$$\vec{A} \cdot \vec{B} = A_x B_x + A_y B_y + A_z B_z. \tag{4.11}$$

The term on the left-hand side of Eq. (4.10) is the four-dimensional dot product, while the term on the right-hand side of Eq. (4.11) is the usual three-dimensional dot product*. Thus, Eqs. (4.10) and (4.11) illustrate one important difference between Euclidean and non-Euclidean geometries. Eq. (4.10) is a definition that will be very useful since the four-dimensional dot product (formally, the total contraction of the tensor indices in the tensor product gAB, as in Eq. (4.13)) is a Lorentz invariant, as we see in Eqs. (4.12).

In the first problem for this chapter, you are asked to show that the four-dimensional dot product is invariant under a Lorentz transformation, so that

$$A' \cdot B' = A \cdot B, \tag{4.12a}$$

or

$$\vec{A'} \cdot \vec{B'} - A'_t B'_t = \vec{A} \cdot \vec{B} - A_t B_t. \tag{4.12b}$$

You can write the four-dimensional dot product in tensor or matrix form; namely,

$$A \cdot B = gAB = \sum_{\mu,\nu} g_{\mu\nu} A_\mu B_\nu, \tag{4.13}$$

---

* Note that we use a dot for both the three- and four-dimensional dot products. Thus, the reader should consider the contexts in order to interpret the meaning of a particular expression involving a dot.

where g is a diagonal matrix (the 1's corresponding to space-like components, the -1 to a time-like component)

$$g = \begin{pmatrix} 1 & 0 & 0 & 0 \\ 0 & 1 & 0 & 0 \\ 0 & 0 & 1 & 0 \\ 0 & 0 & 0 & -1 \end{pmatrix}, \tag{4.14}$$

which is the *metric* for flat SpaceTime in SR. The metric itself is useful in General Relativity, but we will have little use for it in SR. Moreover, in some books the four dimensional dot product is given by

$$A \cdot B = -\vec{A} \cdot \vec{B} + A_t B_t, \tag{4.15}$$

with the diagonal metric

$$g = \begin{pmatrix} -1 & 0 & 0 & 0 \\ 0 & -1 & 0 & 0 \\ 0 & 0 & -1 & 0 \\ 0 & 0 & 0 & 1 \end{pmatrix}. \tag{4.16}$$

Eq. (4.15) is the negative of the dot product in Eq. (4.10). However, since the negative of an invariant quantity is also an invariant, it doesn't really matter.

4.3. Time-like, Space-like, and Null-like Four Vectors and Trajectories

From Eq. (4.10) let $B = A$, so that

$$A \cdot A = \vec{A} \cdot \vec{A} - A_t^2, \tag{4.17}$$

which must also be a Lorentz invariant (i.e.: invariant under a Lorentz transformation). Of course, the sign of the dot product in Eq. (4.17) is also an invariant and tells us whether the four vector is time-like, space-like, or null-like. In particular:

$$\text{If} \begin{cases} A \cdot A < 0, & \text{we say that the four vector is } \textbf{Time-Like} \text{ (since the } A_t^2 \text{ term domi-} \\ & \text{nates)} \\[4pt] A \cdot A > 0, & \text{we say that the four vector is } \textbf{Space-Like} \text{ (since the } \vec{A} \cdot \vec{A} \text{ term dom-} \\ & \text{inates)} \\[4pt] A \cdot A = 0, & \text{we say that the four vector is } \textbf{Null-Like} \text{ (with } \vec{A} \cdot \vec{A} = A_t^2). \end{cases}$$

$$\tag{4.18}$$

These remarks also pertain to SpaceTime differences between events, or four vectors, in which case we can say that the *events* are

$$\begin{cases} \text{Time-Like Separated} \\ \text{Space-Like Separated} \\ \text{Null-Like Separated.} \end{cases} \tag{4.19}$$

Moreover, if we imagine a four vector defining a trajectory in SpaceTime, then—as discussed in subsection 3.4—points in the trajectory are related by *instantaneous Lorentz transformations*, which from the invariance of Eq. (4.17) implies the following. *A given trajectory maintains its time-like, space-like, or null-like character for each of its points in SpaceTime.* In other words, there is no mechanism for switching from, say, a time-like trajectory to a space-like trajectory.

Null-like trajectories are sometimes called null trajectories, the prototypes being the paths of photons (for which $\vec{A} \cdot \vec{A} = A_t^2$). See subsection 4.4.

Recall that we said in subsection 4.1 that the choice of metric does not matter. Well, it does matter in one sense. We are using the metric described by Eq. (4.14). If, however, we use the metric described by Eq. (4.16), the change in sign of the four-dimensional dot product will reverse the signs of the definitions of the space-like and time-like four vectors.

4.4. Proper Length and Proper Time

From Eq. (4.1a), we can construct the differential four vector

$$\mathrm{d}X = (\mathrm{d}x, \mathrm{d}y, \mathrm{d}z, c\,\mathrm{d}t) = (\mathrm{d}\vec{r}, c\,\mathrm{d}t). \tag{4.20}$$

Thus, its four-dimensional dot product must be an invariant; namely,

$$\mathrm{d}X \cdot \mathrm{d}X = \mathrm{d}\vec{r} \cdot \mathrm{d}\vec{r} - c^2\,\mathrm{d}t^2, \tag{4.21}$$

which we will define to be the square of the *differential* of the *Proper Length*, s:

$$\mathrm{d}s^2 = \mathrm{d}X \cdot \mathrm{d}X = \mathrm{d}\vec{r} \cdot \mathrm{d}\vec{r} - c^2\,\mathrm{d}t^2. \tag{4.22}$$

Recall too from Eq. (3.18) the relation for the *differential* of *Proper Time*, τ:

$$\mathrm{d}\tau^2 = \mathrm{d}t^2 - \frac{1}{c^2}\,\mathrm{d}\vec{r} \cdot \mathrm{d}\vec{r}. \tag{4.23}$$

Thus we see that

$$c^2\,\mathrm{d}\tau^2 = -\mathrm{d}s^2. \tag{4.24}$$

Both proper length and proper time are Lorentz invariants, and their differentials are the negative of one another.

Moreover, recall that Eq. (4.23) can be rewritten as (see Eq. (3.20))

$$c^2\,\mathrm{d}\tau^2 = c^2\,\mathrm{d}t^2 - v^2\,\mathrm{d}t^2 = \mathrm{d}t^2(c^2 - v^2), \tag{4.25}$$

with

$$\vec{v} = \frac{\mathrm{d}\vec{r}}{\mathrm{d}t}; \quad v = |\vec{v}|. \tag{4.26}$$

In Eq. (4.25), if $v = c$ for, say, a photon, then $\mathrm{d}\tau = 0$, which is a null trajectory. (Note: photons and other massless particles always travel at the speed of light; with $v = c$.)

4.5. Light Cones and World Lines

We see from subsection 4.3 that time-like and space-like regions of SpaceTime are separated by a null-like boundary. In spatial two dimensions, the boundary is the projection of a three-dimensional *Light Cone*, as we see by comparing Figs. 4.1(a) and 4.1(b). The slopes of the null line ($\tau = 0$) on a two-dimensional slice (projection) are $\pm 45°$, corresponding to c, the speed of light. Such light cones are shown in Figs. 4.1 and 4.2. The point P in Figs. 4.1 and 4.2 can be any point on a *World Line*, which is a trajectory of a particle moving in SpaceTime. Thus, you can imagine a sequence of infinitesimal light cones as we follow a world line through SpaceTime. Also, note from the pictures in Figs. 4.1 and 4.2 that the point P has a *future* light cone (shaded) *above* it and a *past* light cone (shaded) *below* it. These two cones contain all of the points that are causally related to P, whereas the unshaded points outside of the two cones are not causally related to P.

Note that the time-like particle's path, shown in Fig. 4.2(b), can not touch or cross the light cone. Otherwise, it would violate the invariance condition discussed below Eq. (4.19).

Fig. 4.3 shows the upper light cone in Fig. 4.1a and points P, A, B, and C. We see from the geometry that

$$\tan \Theta > 1 \tag{4.27}$$

and

$$\tan \Theta = \frac{c(t_A - t_p)}{x_A - x_B} > 1, \tag{4.28}$$

which implies that

$$c(t_A - t_P) > (x_A - x_P). \tag{4.29}$$

Thus,

$$(x_A - x_P)^2 - c^2(t_A - t_P)^2 < 0, \tag{4.30}$$

so that A and P are time-like separated events. Also, if we let $A \to P$

$$\left(\frac{\mathrm{d}x}{\mathrm{d}t}\right)_A^2 < c^2, \tag{4.31}$$

so that the velocity inside the cone corresponds to a particle whose physical speed does *not* exceed the speed of light. We see then that

> *All physical particles (with $v < c$) follow time-like paths within the past and future light cones.*

Similarly, in Fig. 4.3 it can be easily shown that

$$|c(t_B - t_P)| < |(x_B - x_P)| \tag{4.32}$$

and

$$(x_B - x_P)^2 - c^2(t_B - t_P)^2 > 0. \tag{4.33}$$

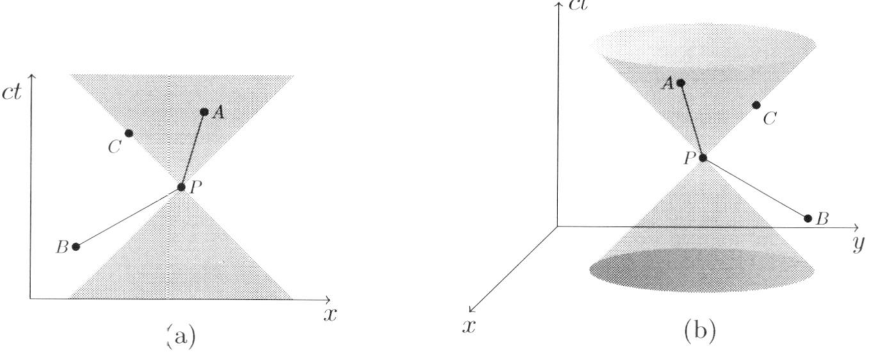

FIGURE 4.1. Light Cones. The right-hand picture (b) shows a two-spatial dimensional light cone for an arbitrary point P, while the left-hand picture (a) is a two-dimensional slice (or projection) of the cone, showing a plot in $ct - x$ space. The projected boundaries of the cone in (a) are the null lines which emanate from P at $\pm 45°$. Point A is time-like separated from P, point B is space-like separated from P, and point C is null-like separated from P. Also, all points in the shaded regions in (a) and (b) are time-like separated from P, while points outside of the shaded regions are space-like separated, with the boundary being the light cone. The light cone is the locus of points that are traced out by a pulse of light emanating from or converging on P. In three spatial dimensions, the pulse of light would be a sphere.

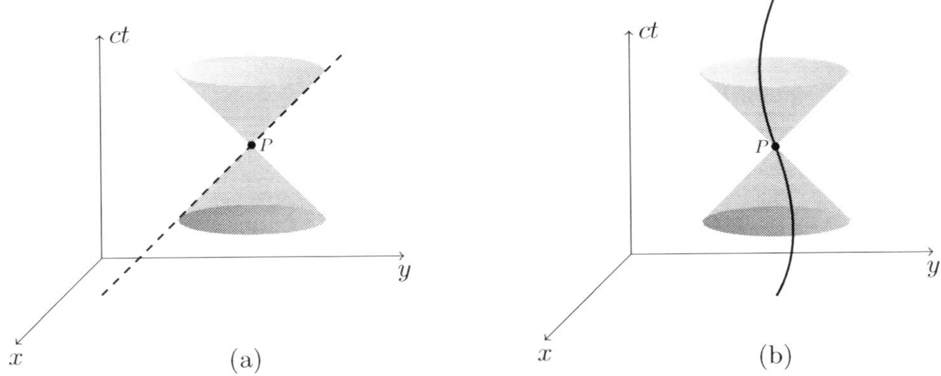

FIGURE 4.2. Trajectory of particles on or within the light cone. (a) shows the path of a photon, whose trajectory is tangent to the light cone. (b) shows a time-like particle's path, which must lie totally within the light cones (future and past).

Moreover, we have as $B \to P$

$$\left(\frac{\mathrm{d}x}{\mathrm{d}t}\right)_B^2 > c^2. \tag{4.34}$$

Therefore, points outside of the past and future light cones correspond to space-like events and particles in this region have speeds greater than that of light, which behavior contradicts SR. So we see that

> *Points outside of the past and future light cones correspond to space-like events and form a region in which no physical particle can exist ($v > c$).*

Finally, for point C on the light cone

$$|c(t_C - t_P)| = |(x_C - x_P)| \tag{4.35}$$

$$(x_C - x_P)^2 - c^2(t_C - t_P)^2 = 0 \tag{4.36}$$

and as $C \to P$

$$\left(\frac{\mathrm{d}x}{\mathrm{d}t}\right)_C^2 \to 1, \tag{4.37}$$

so that

> *The path of a light ray (for a photon) must be tangent to the light cone at every point along its trajectory.*

For such a light ray,

$$\mathrm{d}\tau = 0, \tag{4.38}$$

which pertains to every point on the surface of the past and future light cones.

Moreover, the velocities we have derived for points A, B, and C were taken to be in the limit as the point $\to P$. However, these velocities could be considered to pertain to straight lines between the point P and points A, B, and C. Then, it is clear that if you deform the path so that it is no longer a straight line, because of the invariance mentioned before, you can not change its space-like or time-like nature, so we have obtained general characteristics of the points within, on and outside of the light cones.

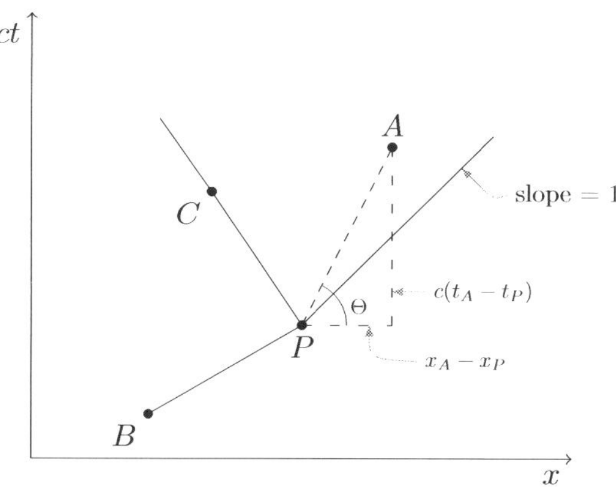

FIGURE 4.3. Upper light cone, showing points P, A, B, and C in Fig. 4.1a

4.6. Causality

See a typical space-like event B on the $ct - x$ plot shown in Fig. 4.4. We see that in the unprimed system event P causes event B, so that

$$ct_B > ct_P \qquad (\text{or } t_B > t_P). \tag{4.39}$$

However, it is clear that we can Lorentz transform to a coordinate system (intertial frame, given by ct' and x' in Fig. 4.4), in which

$$ct'_B < ct'_P \qquad (\text{or } t'_B < t'_P). \tag{4.40}$$

This can never be accomplished with time-like separated events, so that

The order in which events are seen can be reversed for space-like events,

which means that

Causality can be violated for space-like separated events, corresponding to non-physical particles which have speeds greater than the speed of light.

Consider now increments for the Lorentz equation given in Eq. (4.2d) (with $x_1 = x$)

$$c\Delta t' = \gamma \left(c\Delta t - \frac{u}{c}\Delta x \right). \tag{4.41}$$

If $\Delta t > 0$, under what conditions will $\Delta t' < 0$? From Eq. (4.41), this occurs when

$$c\Delta t < \frac{u}{c}\Delta x, \tag{4.42}$$

or, since $u < c$

$$c\Delta t < \Delta x, \tag{4.43}$$

or taking differentials:

$$\frac{\mathrm{d}x}{\mathrm{d}t} > c. \tag{4.44}$$

We see again that you can transform to an inertial system in which causality is violated for space-like separated events corresponding to particles with speeds greater than that of light.

At one time physicists were very much interested in the possible existence of tachyons, which are elementary particles that move faster than the speed of light. Experiments were performed to

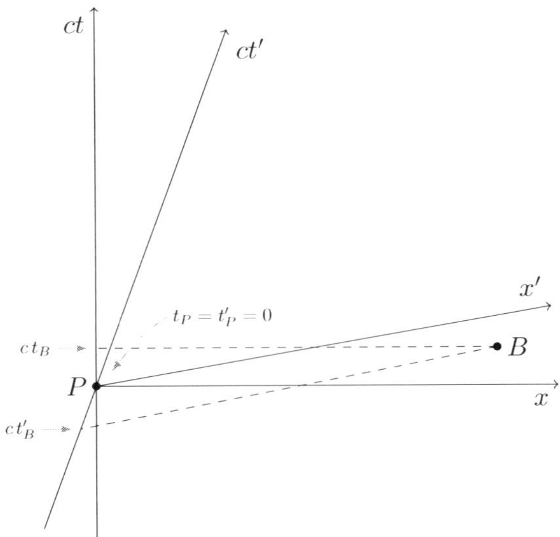

FIGURE 4.4. Behavior of two space-like separated events P and B on a $ct - x$ plot showing the $ct' - x'$ lines of a Lorentz transformation which reverses the order of the events.

verify their existence, the results of which were all negative. Thus, no elementary particles (or other objects) can move faster than the speed of light. Tachyons have been immortalized in this famous limerick by Reginald Buller:

> There was a young lady named Bright,
> Whose speed was far faster than light.
> > She went out one day,
> > In a relative way,
> And returned the previous night!

4.7. Time Travel

Thus, SR has implications for time travel. From the resolution of the twin paradox discussed in Chapter 3, it is clear that time travel into the future is possible. But, in order to obtain a significant time difference you would need to travel in a spaceship averaging a speed that is a fair fraction of the speed of light.

However, in SR time travel into the past is not allowed since such a happening would involve space-like events, with speeds greater than c, for which causality is violated. It is indeed fortunate that time travel into the past is forbidden by SR; otherwise, it would be possible to travel there and murder one or both of your parents, which is a logical impossibility.

It has been claimed that time travel into the past might be possible using wormholes, but wormholes are highly speculative and still leave us with the causality dilemma. Wormholes are allowed by General Relativity, but there appears to be no current empirical evidence for their existence, especially for wormholes of macroscopic size. Also, Stephen Hawking has proposed his famous *Chronology Protection Conjecture*, which states that "The laws of physics conspire to prevent time travel by macroscopic objects". See Reference 53, pg. 521.

4.8. The Four-Gradient, The Four-Divergence, and the D'Alembertian Operator

In classical physics, it is useful to analyze the behavior with vector calculus. This analysis involves the ∇ operator, applications of which include the following: gradient $\phi = \nabla\phi$, a vector, giving the spatial change of a scalar function ϕ, while $\nabla \cdot \vec{A}$ is the divergence of a vector $\vec{A}$, which gives the behavior of small circles (e.g., in fluid flow) around a source or sink. We want to generalize this mathematics to relativity.

Define a scalar function $\mathcal{X}$ that depends on x and t (or x and ct)

$$\mathcal{X} = \mathcal{X}(x, ct). \tag{4.45}$$

Suppose we consider a small change Δt in $\mathcal{X}$ while holding x constant ($\Delta x = 0$), for which we obtain for an observer in S

$$\Delta\mathcal{X} = \frac{\partial\mathcal{X}}{\partial(ct)}\Delta(ct). \tag{4.46}$$

However, according to the moving observer in S', we have

$$\Delta\mathcal{X} = \frac{\partial\mathcal{X}}{\partial x'}\Delta x' + \frac{\partial\mathcal{X}}{\partial(ct')}\Delta(ct'). \tag{4.47}$$

From Eqs. (4.2a) and (4.2d), we note that

$$\Delta x' = \gamma\left(-\frac{u}{c}\Delta(ct)\right) \tag{4.48}$$

$$c\Delta t' = \gamma\Delta(ct),$$

where we recall that $\gamma = \frac{1}{\sqrt{1-u^2/c^2}}$ and $\Delta x = 0$. Then, from Eqs. (4.46), (4.47), and (4.48) we obtain

$$\Delta\mathcal{X} = \gamma\left[-\frac{u}{c}\frac{\partial\mathcal{X}}{\partial x'} + \frac{\partial\mathcal{X}}{\partial(ct')}\right]\Delta(ct) = \frac{\partial\mathcal{X}}{\partial(ct)}\Delta(ct), \tag{4.49}$$

where the last line follows from the fact that $\mathcal{X}$ is a scalar function, which is the same in any inertial frame. From Eq. (4.49) we find that

$$\frac{\partial\mathcal{X}}{\partial(ct)} = \gamma\left(\frac{\partial\mathcal{X}}{\partial(ct')} - \frac{u}{c}\frac{\partial\mathcal{X}}{\partial x'}\right). \tag{4.50}$$

Next, consider an analogous derivation for which there is a small change in x, Δx, keeping the time t constant ($\Delta t = 0$), so that

$$\frac{\partial\mathcal{X}}{\partial x} = \gamma\left(\frac{\partial\mathcal{X}}{\partial x'} - \frac{u}{c}\frac{\partial\mathcal{X}}{\partial(ct')}\right). \tag{4.51}$$

If we compare Eqs. (4.50) and (4.51) with the inverse transformation to Eqs. (4.2), we see that the appropriate four vector for the four gradient is *either*

$$\partial = \left(-\vec{\nabla}, \frac{\partial}{\partial(ct)}\right), \tag{4.52}$$

or

$$\partial = \left(\vec{\nabla}, -\frac{\partial}{\partial(ct)}\right). \tag{4.53}$$

We take the latter, Eq. (4.53), to be the *four-gradient*.

Next, define the four-divergence as

$$\partial \cdot A = \vec{\nabla} \cdot \vec{A} - \left(-\frac{\partial}{\partial(ct)}\right)A_t$$

or

$$\partial \cdot A = \vec{\nabla} \cdot \vec{A} + \frac{\partial}{\partial(ct)} A_t. \tag{4.54}$$

Finally, we generalize the three-dimensional Laplacian operator

$$\nabla^2 = \vec{\nabla} \cdot \vec{\nabla} = \frac{\partial^2}{\partial x^2} + \frac{\partial^2}{\partial y^2} + \frac{\partial^2}{\partial z^2} \tag{4.55}$$

to four dimensions, obtaining

$$\partial \cdot \partial = \vec{\nabla} \cdot \vec{\nabla} - \left(-\frac{\partial}{\partial(ct)}, -\frac{\partial}{\partial(ct)} \right) \tag{4.56}$$

$$= \nabla^2 - \left(\frac{\partial}{\partial(ct)} \right)^2,$$

which is usually written

$$\Box^2 = \nabla^2 - \frac{\partial^2}{\partial(ct)^2}, \tag{4.57}$$

where $\Box^2$ is the d'Alembertian operator.

Applications of this relativistic vector calculus will be explored in Chapter 9 on electrodynamics. Also, in Quantum Mechanics the operator form of the four vector momentum is given by $P = -i\hbar\partial = -i\hbar(\nabla, -\frac{1}{c}\frac{\partial}{\partial t})$. as we discuss in Chapter 12 for the Dirac equation.

4.9. Problems

(1) Consider two four vectors $A = (\vec{A}, A_t)$ and $A' = (\vec{A}', A'_t)$, each evaluated in two inertial frames of reference S and S', respectively, which are *related by a generalized Lorentz transformation* (i.e., that pertaining to the four vector, A, which has exactly the same form as the transformation relating space and time), i.e.,

$$A'_x = \gamma \left[A_x - \frac{u}{c} A_t \right]$$

$$A'_y = A_y \quad ; \quad A'_z = A_z$$

$$A'_t = \gamma \left[A_t - \frac{u}{c} A_x \right]$$

where

$$\gamma = \frac{1}{\sqrt{\left[1 - \left(\frac{u}{c}\right)^2 \right]}}.$$

Then, consider two four vectors $B = (\vec{B}, B_t)$ and $B' = (\vec{B}', B'_t)$, each evaluated in S and S', respectively, which are also *related by the same generalized Lorentz transformation*; i.e.,

$$B'_x = \gamma \left[B_x - \frac{u}{c} B_t \right]$$

$$B'_y = B_y \quad ; \quad B'_z = B_z$$

$$B'_t = \gamma \left[B_t - \frac{u}{c} B_x \right].$$

Show that

$$-A_t B_t + \vec{A} \cdot \vec{B} = -A'_t B'_t + \vec{A}' \cdot \vec{B}',$$

or

$$A \cdot B = A' \cdot B'.$$

Thus, $A \cdot B = -A_t B_t + \vec{A} \cdot \vec{B}$ is a *relativistically invariant quantity*. Note:

$$\vec{A} \cdot \vec{B} = A_x B_x + A_y B_y + A_z B_z$$

is a *three-dimensional dot product*. Also, the *y and z components of the four vectors do not change* in the generalized Lorentz transformation. Thus the y and z contributions concel out on both sides of the equations, so one need only prove that

$$-A_t B_t + A_x B_x = -A'_t B'_t + A'_x B'_x.$$

(This is the equation you have to prove.)

It is also clear that for a single four vector, the invariant quantity is $-(A_t)^2 + \vec{A} \cdot \vec{A}$, which we see from above by setting $B = A$.

Also, note the convenient and useful identity:

$$\gamma^2 \left[1 - \left(\frac{u}{c}\right)^2 \right] = 1.$$

(2) Consider two *four vectors*, whose components are given by:

$$A_\mu : (-1, 0, 1, 2);$$

$$B_\mu : (4, 0, 2, 2).$$

37

(a) State, for each of these vectors, whether it is *time-like*, *space-like*, or *null-like*.
(b) Find the following:
     (i) $A - 5B$
     (ii) $A - B$
(c) Calculate the four-dimensional *dot product* between A and B.

(3) Consider the three four-vectors whose components are given by:

$$A_\mu : (5, 0, 2, 1);$$
$$B_\mu : (3, 0, 4, 5);$$
$$C_\mu : (6, 0, 8, 7).$$

(a) State, for each of these vectors, whether it is *time-like*, *space-like*, or *null-like*.
(b) Find the following:
     (i) $A + B + C$
     (ii) $A - 2B - 3C$
     (iii) $5A$
(c) Calculate the four-dimensional *dot product* between A and B.
(d) Calculate the four-dimensional *dot product* between A and C.
(e) Are the four-dimensional dot products in parts (c) and (d) invariant under a Lorentz transformation?

Chapter **5**

Relativisitic Kinematics and Dynamics; $E = mc^2$

5.1. Constructing Four Vectors

Start with the differential of the four vector in Eq. (4.4), giving

$$\mathrm{d}A = (\mathrm{d}\vec{A}, \mathrm{d}A_t), \tag{5.1}$$

which itself is a four vector, with the invariant

$$\mathrm{d}A \cdot \mathrm{d}A = \mathrm{d}\vec{A} \cdot \mathrm{d}\vec{A} - \mathrm{d}A_t^2. \tag{5.2}$$

Next construct

$$\frac{\mathrm{d}A}{\mathrm{d}\tau} = \left(\frac{\mathrm{d}\vec{A}}{\mathrm{d}\tau}, \frac{\mathrm{d}A_t}{\mathrm{d}\tau} \right), \tag{5.3}$$

where $\mathrm{d}\tau$ is the differential of proper time, defined in Eq. (4.23). The quantity defined in Eq. (5.3) must itself be a Lorentz invariant four vector, with the invariant

$$\frac{\mathrm{d}A}{\mathrm{d}\tau} \cdot \frac{\mathrm{d}A}{\mathrm{d}\tau} = \frac{\mathrm{d}\vec{A} \cdot \mathrm{d}\vec{A}}{(\mathrm{d}\tau)^2} - \left(\frac{\mathrm{d}A_t}{\mathrm{d}\tau} \right)^2. \tag{5.4}$$

Thus, since $\mathrm{d}\tau$ is a Lorentz invariant, if A is a four vector, $\frac{\mathrm{d}A}{\mathrm{d}\tau}$ must also be a four vector, whose four-dimensional dot product in Eq. (5.4) is a Lorentz invariant. Then, as we will see, τ in SR plays an analogous role to t in Newtonian physics.

It will be useful to construct invariants using $1/\mathrm{d}\tau$, which formally means taking derivatives with respect to proper time. First, however, we briefly review Newtonian mechanics.

5.2. Newtonian Kinematics and Dynamics

In classical mechanics we can define and differentiate with respect to the time t, which is an invariant in Galilean physics. See Eq. (2.1d). First, we describe a particle's path by

$$\vec{r} = \vec{r}(t), \tag{5.5}$$

with a velocity given by

$$\vec{v} = \frac{\mathrm{d}\vec{r}}{\mathrm{d}t} = \frac{\mathrm{d}\vec{r}}{\mathrm{d}s} \frac{\mathrm{d}s}{\mathrm{d}t} = v \frac{\mathrm{d}\vec{r}}{\mathrm{d}s}, \tag{5.6}$$

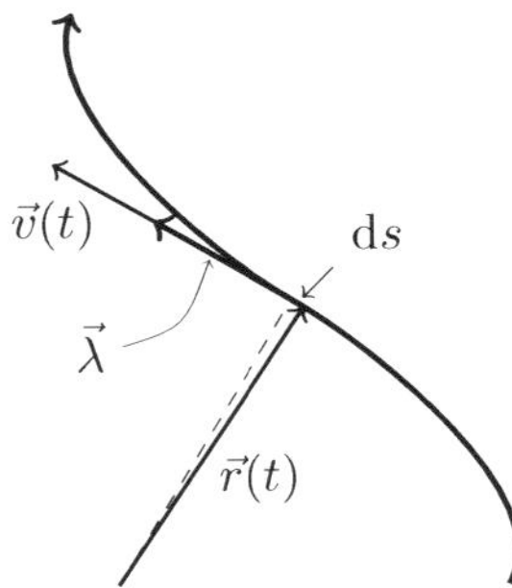

FIGURE 5.1. Classical path described by $\vec{r} = \vec{r}(t)$.

where $\mathrm{d}s$ is the differential element of path length shown in Fig. 5.1, and

$$v = \frac{\mathrm{d}s}{\mathrm{d}t} \tag{5.7}$$

is the infinitesimal speed of the particle along its path. In elementary texts, it is shown that

$$\vec{\lambda} = \frac{\mathrm{d}\vec{r}}{\mathrm{d}s} \tag{5.8}$$

is a *unit vector tangent* to the path. Thus, we obtain

$$\vec{v} = \vec{v}(t) = v\vec{\lambda}. \tag{5.9}$$

See problem 1 for this chapter.

In classical mechanics, it is then convenient to define the acceleration, momentum, and force for a particle:

$$\vec{a} = \vec{a}(t) = \frac{\mathrm{d}\vec{v}(t)}{\mathrm{d}t} = \frac{\mathrm{d}^2\vec{r}(t)}{\mathrm{d}t^2}; \tag{5.10}$$

$$\vec{p} = \vec{p}(t) = m\vec{v}; \tag{5.11}$$

$$\vec{F} = m\vec{a} = m\frac{\mathrm{d}\vec{v}}{\mathrm{d}t} = \frac{\mathrm{d}\vec{p}}{\mathrm{d}t}, \tag{5.12}$$

where m is the mass of the particle. In what follows, we will find analogues in SR for Eqs. (5.9), (5.10), (5.11), and (5.12). However, in SR we can no longer use t as a parameter with which to differentiate since t is no longer an invariant. (This follows really from the Lorentz equation Eqs. (2.6d), where $t' \neq t$). Instead we will use τ which is a relativistic invariant, following the procedure outlined in subsection 5.1.

However, in this procedure we emphasize that the invariance implied by the ratio of $\frac{\mathrm{d}A}{\mathrm{d}\tau}$ has a special meaning. The quantity $\frac{\mathrm{d}A}{\mathrm{d}\tau}$ is not itself constant along a world line for which we have a sequence of infinitestimal inertial frames. Rather, what is invariant is the quantity defined in Eq. (5.4). In the construction of various four vectors, we will evaluate the appropriate Lorentz invariants.

5.3. Kinematics in SR

From the four vector defined in Eq. (4.20) we construct the four vector (as defined by Eq. (5.3)):

$$U = \frac{\mathrm{d}X}{\mathrm{d}\tau} = \left(\frac{\mathrm{d}\vec{r}}{\mathrm{d}\tau},\; c\,\frac{\mathrm{d}t}{\mathrm{d}\tau} \right), \tag{5.13}$$

which should not be confused with u, the frame velocity. Then, from Eq. (4.25), we find that

$$\mathrm{d}\tau = \mathrm{d}t\sqrt{1 - \frac{v^2}{c^2}} = \frac{\mathrm{d}t}{\gamma}, \tag{5.14}$$

with

$$\gamma = \left(1 - \frac{v^2}{c^2}\right)^{-1/2}, \tag{5.15}$$

where we continue with the understanding, mentioned in Chapter 3, that the velocity of a particle

$$\vec{v} = \frac{\mathrm{d}\vec{r}}{\mathrm{d}t} \qquad (v = |\vec{v}|) \tag{5.16}$$

can be considered an instantaneous frame velocity. Thus, $\vec{r}$ and t are also evaluated in the particular instantaneous frame. From Eqs. (5.13) and (5.14) we obtain

$$U = \frac{\mathrm{d}X}{\mathrm{d}\tau} = \gamma(\vec{v}, c), \tag{5.17}$$

which is the four-velocity "tangent" to the curve and is the SR analogue of Eq. (5.9) in Newtonian mechanics.

Next, we find the invariant associated with U:

$$U \cdot U = \gamma^2(\vec{v} \cdot \vec{v} - c^2)$$

$$= -\gamma^2 c^2 \left(1 - \frac{v^2}{c^2}\right)$$

$$= -\gamma^2 c^2 \frac{1}{\gamma^2},$$

or

$$U \cdot U = -c^2, \tag{5.18}$$

a result that can be derived another way using Eqs. (5.17), (4.22), and (4.24); namely,

$$U \cdot U = \frac{\mathrm{d}X \cdot \mathrm{d}X}{\mathrm{d}\tau^2} = \frac{\mathrm{d}s^2}{\mathrm{d}\tau^2} = -c^2.$$

We can also define a *four acceleration a* and *four momentum P*:

$$a = \frac{\mathrm{d}U}{\mathrm{d}\tau} = \frac{\mathrm{d}^2 X}{\mathrm{d}\tau^2} \tag{5.19}$$

$$P = m_\circ U = m_\circ \gamma(\vec{v}, c). \tag{5.20}$$

These are the analogues of the classical equations (5.10) and (5.11). However, we remark that the mass $m_\circ$ in Eq. (5.20) is called the *rest mass* since it is the mass when the particle is at rest (in a frame with $v = 0$). In SR the mass can depend on velocity. (See Eq. (5.27).)

An important result is derived in Problem 2; namely we find that

$$a \cdot U = 0, \tag{5.21}$$

so that the four velocity and acceleration are *normal,* or *orthogonal,* to one another in SpaceTime. (This behavior is the same as the velocity and acceleration, in Newtonian physics, for a particle moving in a circle.)

5.4. The Most Famous Equation in Physics

We next rewrite Eq. (5.20) as follows

$$P = \left(\vec{p}, \frac{E}{c} \right), \tag{5.22}$$

where

$$\vec{p} = m_\circ \gamma \vec{v}, \tag{5.23}$$

and

$$\frac{E}{c} = m_\circ \gamma c, \tag{5.24}$$

where γ is defined in Eq. (5.15). The time-like component of the four-momentum is given by Eq. (5.24) and by definition is the $\frac{\text{energy}}{c} = \frac{E}{c}$. Notice too that $\vec{p}$ and $\frac{E}{c}$ have the same units. Eq. (5.24) can then be reexpressed as

$$E = m_\circ \gamma c^2 = \frac{m_\circ c^2}{\left(1 - \frac{v^2}{c^2}\right)^{1/2}}, \tag{5.25}$$

or

$$E = mc^2, \tag{5.26}$$

with the velocity-dependent mass given by

$$m = \frac{m_\circ}{\sqrt{1 - \frac{v^2}{c^2}}}. \tag{5.27}$$

Eq. (5.26) is the most famous equation in physics. What the layman doesn't realize is that the mass in the equation depends on the velocity, as in Eq. (5.27). However, the particle strongly "resists" having a velocity $|\vec{v}| \to c$. Just as in Newtonian physics, as the velocity increases we increase both momentum and the energy and

$$\lim_{v \to c} (\vec{p}, E) \to \infty. \tag{5.28}$$

Next, go to the classical (Newtonian) limit defined by

$$\left| \frac{v}{c} \right| \ll 1, \tag{5.29}$$

so that

$$\left(1 - \frac{v^2}{c^2}\right)^{-1/2} = 1 + \frac{1}{2}\frac{v^2}{c^2} + \dots, \tag{5.30}$$

showing explicitly only the first two terms of the Taylor expansion. Then, from Eq. (5.25)

$$E = m_\circ c^2 + \frac{1}{2} m_\circ v^2 + \dots \tag{5.31}$$

The second term in Eq. (5.31) is the classical kinetic energy, while the first term, which has no classical analogue, is called the *rest mass energy*

$$E_\circ = m_\circ c^2. \tag{5.32}$$

We now write in general

$$E = E_\circ + T, \tag{5.33}$$

where T is the relativistic kinetic energy, so that

$$T = E - m_\circ c^2 = m_\circ c^2 \left[\left(1 - \frac{v^2}{c^2} \right)^{-1/2} - 1 \right]. \tag{5.34}$$

The relativistic spatial part of the four-momentum is given by

$$\vec{p} = m_\circ \gamma \vec{v} = \frac{m_\circ}{\sqrt{1 - \frac{v^2}{c^2}}} \, \vec{v}. \tag{5.35}$$

As described in Problem 3, we find that the invariant associated with the four-momentum in Eq. (5.22) is given by

$$P \cdot P = \vec{p} \cdot \vec{p} - \frac{E^2}{c^2} = -m_\circ^2 c^2, \tag{5.36}$$

or

$$E^2 = p^2 c^2 + m_\circ^2 c^4, \tag{5.37}$$

with

$$p = |\vec{p}| = (\vec{p} \cdot \vec{p})^{1/2}. \tag{5.38}$$

Eq. (5.36) tells us that: (i) the rest mass $m_\circ$ is a relativistically invariant quantity and (ii) energy and momentum get "mixed up" in going from one infinitesimal inertial frame to another but in any frame there is the invariant given by Eq (5.36).*

From Eqs. (5.16), (5.25), (5.35), and (5.38) we find that

$$pc = \frac{Ev}{c}, \tag{5.39}$$

where we note that v and p are the absolute values of $\vec{v}$ and $\vec{p}$, respectively.

5.5. Photons

Photons, or other massless particles, have

$$m_\circ = 0, \quad v = |\vec{v}| = c, \tag{5.40}$$

which gives indeterminate expressions for $\vec{p}$ in Eq. (5.35) and E in Eq. (5.25). However, in Eq. (5.37), if we let $m_\circ \to 0$

$$E^2 = p^2 c^2$$

or†

$$E = pc, \tag{5.41}$$

which also follows from Eq. (5.39) by letting $v \to c$ and which also justifies the choice of the $+$ sign in Eq. (5.41).

Thus, photons have both energy and momentum, which are related through Eq. (5.41).

Other particles which are thought to be massless are the gluon (which is confined to hadrons and glueballs) and the graviton (if it exists). Moreover, it was once thought that neutrinos are massless, in which case the above formalism would have applied to neutrinos. We now know that neutrinos have tiny but non-zero masses.

---

*However, we shall show in the next chapter that energy and momentum are separately conserved in a given inertial frame.

†Really formally $E = \pm pc$.

Quantity	Classical Expression	SR (four quantities)
Velocity	$\vec{v} = \dfrac{d\vec{r}}{dt}$	$U = \dfrac{dx}{d\tau} = \gamma(\vec{v}, c)$
Acceleration	$\vec{a} = \dfrac{d\vec{v}}{dt} = \dfrac{d^2\vec{r}}{dt^2}$	$a = \dfrac{dU}{d\tau} = \dfrac{d^2 x}{d\tau^2}$
Momentum	$\vec{p} = m\vec{v} = m\dfrac{d\vec{r}}{dt}$	$P = m_\circ U = (\vec{p}, \frac{E}{c})$ $\vec{p} = \gamma m_\circ \vec{v}$
Energy (no potential energy)	$E = T = \frac{1}{2}mv^2$	$E = m_\circ c^2 + T$ $= \gamma m_\circ c^2 = \dfrac{m_\circ c^2}{\sqrt{1 - \frac{v^2}{c^2}}}$
Force	$\vec{F} = \dfrac{d\vec{p}}{dt}$	$F = m_\circ a = \dfrac{dP}{d\tau} = \gamma(\vec{F}, \frac{1}{c}\vec{v} \cdot \vec{F})$ $\vec{F} = \dfrac{d\vec{p}}{dt} = \dfrac{d}{dt}(\gamma m_\circ \vec{v})$

TABLE 5.1. Comparisons between analogous quantities in Classical (Newtonian) Mechanics and SR.

5.6. Dynamics in SR

We can define a *four-force* in SR, in analogy with the classical Eq. (5.12). We accomplish this by differentiating the four momentum in Eq. (5.22) with respect to proper time, so that

$$F = \frac{dP}{d\tau} = m_\circ \frac{dU}{d\tau}$$
$$= \frac{d}{d\tau}\left(\vec{p}, \frac{E}{c}\right). \tag{5.42}$$

Then, from Eq. (5.14), Eq. (5.42) becomes

$$F = \frac{dP}{d\tau} = \gamma\left(\frac{d\vec{p}}{dt}, \frac{1}{c}\frac{dE}{dt}\right), \tag{5.43}$$

which is the *four-force* (sometimes called the *Minkowski force*). As you see, the four-force gives the rate of change of four-momentum. Its spatial part is the rate of change of spatial momentum and its time-like part gives the rate of change of energy.

Next, differentiate Eq. (5.37) to obtain

$$2E\frac{dE}{dt} = 2c^2\vec{p} \cdot \frac{d\vec{p}}{dt}, \tag{5.44}$$

or since $\vec{p} = \gamma m_\circ \vec{v}$ and $E = \gamma m_\circ c^2$

$$\frac{dE}{dt} = \vec{v} \cdot \frac{d\vec{p}}{dt}. \tag{5.45}$$

We also define

$$\vec{F} = \frac{d\vec{p}}{dt}, \tag{5.46}$$

which is not the same as the classical expression, Eq. (5.12) since $\vec{p}$ contains a factor of γ. Substituting Eqs (5.45) and (5.46) into Eq. (5.43) we find that

$$F = \gamma\left(\vec{F}, \frac{1}{c}\vec{v} \cdot \vec{F}\right). \tag{5.47}$$

Note too that the time-like part of F is similar to the classical expression for the rate of change of energy:

$$\frac{\mathrm{d}E}{\mathrm{d}t} = \frac{\vec{F} \cdot \mathrm{d}x}{\mathrm{d}t} = \vec{F} \cdot \vec{v}.$$

However, again because of the factor of γ in Eq. (5.47), we should not overly emphasize the analogies with the classical expression.

Also, there is again an invariant

$$\begin{aligned}
F \cdot F &= \gamma^2 \left[\vec{F} \cdot \vec{F} - \frac{1}{c^2} (\vec{v} \cdot \vec{F})^2 \right] \\
&= \gamma^2 \vec{F} \cdot \left[\vec{F} - \frac{1}{c^2} (\vec{v} \cdot \vec{F}) \, \vec{v} \right],
\end{aligned} \tag{5.48}$$

which is not especially interesting or useful.

Finally, we summarize the differences between the classical and relativistic expressions in Table 5.1.

5.7. Problems

(1) For a curve in classical mechanics described by the equation, $\vec{r} = \vec{r}(t)$, prove that

$$\vec{v} = \vec{v}(t) = \frac{\mathrm{d}\vec{r}}{\mathrm{d}t} = v\vec{\lambda},$$

where $v = \frac{\mathrm{d}s}{\mathrm{d}t}$ is the scalar speed along the path and

$$\vec{\lambda} = \frac{\mathrm{d}\vec{r}}{\mathrm{d}s}$$

is a **unit vector** *tangent* to the path. For the last part of this proof, draw a picture.

(2) From the expression derived in the text for the **four-velocity**, U, and the **four-acceleration**, a, show that the **four-dimensional dot product** of U and a is equal to zero, i.e.,

$$-U_t a_t + \vec{U} \cdot \vec{a} = 0.$$

Thus the relative four-acceleration is "normal" or orthogonal to the relativistic four-velocity. Hint: the fastest way to do this problem is to use Eq. (5.18).

(3) From the definitions of energy and momentum, find the **relativistically invariant quantity** for the momentum/energy four-vector $(\vec{p}, E/c)$. (See Problem #1, Chapter 4). Note: you can *assume*, if you like, that the particle moves in the x-**direction**. In that case, the magnitude of the velocity in the x-direction is the same as the magnitude of the total velocity, $v = |\vec{v}| = |v_x|$. Similarly, the magnitude of the momentum in the x-direction is the same as the magnitude of the total momentum. We have

$$E = \gamma m_\circ c^2 \quad ; \quad p = \gamma m_\circ v,$$

where

$$\gamma = \frac{1}{\sqrt{1 - \left(\frac{v}{c}\right)^2}},$$

and $m_\circ$ is the rest mass. Complete the proof by inserting the above quantities into the four-dimensional dot product:

$$P \cdot P = -\frac{E^2}{c^2} + \vec{p} \cdot \vec{p} = -\frac{E^2}{c^2} + p^2,$$

and show that:

$$P \cdot P = -m_\circ^2 c^2.$$

(4) It will be useful to derive two identities in order to complete a proof:

(a) With the definition given in Eq. (5.15), $\gamma = \left(1 - \left(\frac{v}{c}\right)^2\right)^{-1/2}$, prove that

$$\frac{\mathrm{d}\gamma}{\mathrm{d}t} = \frac{\gamma^3}{c^2}\left(\vec{v} \cdot \frac{\mathrm{d}\vec{v}}{\mathrm{d}t}\right).$$

Recall that $v^2 = \vec{v} \cdot \vec{v}$.

(b) Show that

$$\gamma^2 \frac{v^2}{c^2} + 1 = \gamma^2.$$

(c) Using the equations

$$F = m_\circ \left(\frac{\mathrm{d}U}{\mathrm{d}\tau} \right) = m_\circ \frac{\mathrm{d}}{\mathrm{d}\tau}[\gamma(\vec{v}, c)],$$

and

$$\frac{1}{\mathrm{d}\tau} = \frac{\gamma}{\mathrm{d}t},$$

take the indicated derivitives and show that you get the same result as in the text; namely,

$$F = \frac{\mathrm{d}P}{\mathrm{d}\tau} = \gamma \left(\vec{F}, \frac{\vec{v} \cdot \vec{F}}{c} \right), \qquad \vec{F} = \frac{\mathrm{d}\vec{p}}{\mathrm{d}t}; \ \vec{p} = \gamma m_\circ \vec{v};$$

where you should find it helpful to use the identities derived in parts (a) and (b).

(5) A $\pi_\circ$ meson has a rest mass of 135 MeV, which really means 135 MeV/c^2, so that the rest mass-energy is 135 MeV. Assume that this particle has a magnitude of speed of $v = \frac{c}{\sqrt{2}}$ in a direction of 45° to the x-axis, where we also assume that the motion is in the x-y plane, so that the x- and y-components of the four-velocity are the same.

(a) Find the components of the four-velocity of the particle.

(b) Find the components of the energy/momentum four-vector.

It is convenient to express the components of the four-velocity in units of c, and the energy/momentum four-vector in units of MeV/c.

If I had but known, I would have been a locksmith.

Albert Einstein

Chapter 6

The Principle of Extremal Aging

6.1. The Calculus of Variations

The calculus of variations is developed in many books. Thus, we will not derive the formalism here, but merely quote the main results. However we do assert that the crux of the method is to find the extremum of a so-called functional, which is sometimes called a function of a function or a function whose space consists of a family of functions. A functional, which provides a mechanism for using a function to give a number, is illustrated by Eq. (6.1) in which the functional S depends upon the function of x. Then, we vary the form of x (giving a path) to extremize S. Exterema of functionals are found to generate fundamental, important equations or principles in science, mathematics, and engineering.

In classical mechanics we have the *Principle of Extremal Action,* leading to *Lagrange's Equations of Motion.* Consider first the one-dimensional spatial case shown in Fig. 6.1, with $x = x(t)$. There are an infinite number of paths between points A and B. We want to find the path (or paths) by finding an extremum of the *functional* defined by

$$\text{Action} = \mathcal{S}[x(t)] = \int_{t_A}^{t_B} \mathrm{d}t \; \mathcal{L}(\dot{x}(t), x(t)), \tag{6.1}$$

where t is the time, $\dot{x}$ is the time derivative of x, and the brackets for $\mathcal{S}$ denote path dependence, and $\mathcal{L}$ is the Lagrangian function. Note that, for each path in space, we obtain a number for $\mathcal{S}$, and we want to find the path (or paths) for which that number is an extremum. The procedure for the functional $\mathcal{S}$ is different from the usual extremum methods in calculus, for which we set $\frac{\mathrm{d}F(x)}{\mathrm{d}x} = 0$ to obtain extrema for a function $F(x)$.

The variational equation one obtains is then given by

$$\frac{\mathrm{d}}{\mathrm{d}t}\left(\frac{\partial \mathcal{L}}{\partial \dot{x}}\right) - \frac{\partial \mathcal{L}}{\partial x} = 0. \tag{6.2}$$

In classical mechanics one defines the Lagrangian as

$$\mathcal{L}(\dot{x}, x) = T(\dot{x}, x) - V(x), \tag{6.3}$$

49

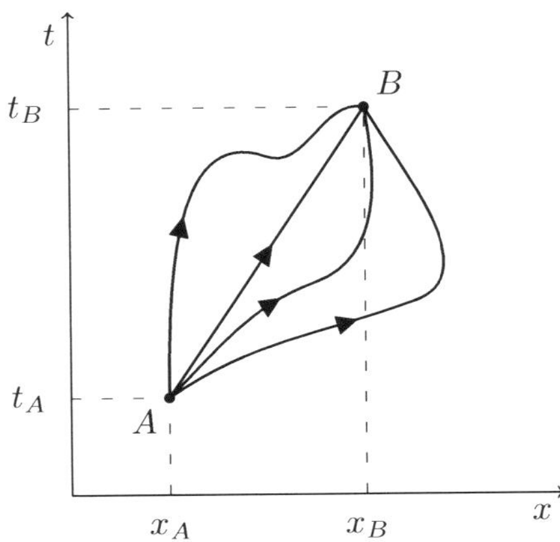

FIGURE 6.1. Paths between two SpaceTime points (x_A, t_A) and (x_B, t_B). An infinite number of paths is possible, but normally only one path satisfies the variational principle.

where T is the kinetic energy function and V is the potential energy function. Thus, for the simple case of a particle in a potential V

$$\mathcal{L}(\dot{x}, x) = \frac{1}{2} m \dot{x}^2 - V(x),\tag{6.4}$$

we obtain from Eq. (6.2)

$$\frac{\mathrm{d}}{\mathrm{d}t}(m\dot{x}) + \frac{\mathrm{d}V}{\mathrm{d}x} = 0,$$

or

$$m\ddot{x} = -\frac{\mathrm{d}V}{\mathrm{d}x},\tag{6.5}$$

which is Newton's Second Law in the presence of a potential. For free-particle motion, $V = 0$ and

$$m\ddot{x} = 0.\tag{6.6}$$

In classical mechanics, we may have a system of N generalized coordinates

$$q_1, q_2, \ldots, q_N$$

and their N generalized velocities

$$\dot{q}_1, \dot{q}_2, \ldots, \dot{q}_N,$$

so that the Lagrangian an be written as

$$\mathcal{L} = \mathcal{L}(q_1, q_2, \ldots, q_N, \dot{q}_1, \dot{q}_2, \ldots, \dot{q}_N) = T - V.\tag{6.7}$$

The generalization of Eq. (6.2) is

$$\frac{\mathrm{d}}{\mathrm{d}t}\left(\frac{\partial \mathcal{L}}{\partial \dot{q}_\alpha}\right) - \left(\frac{\partial \mathcal{L}}{\partial q_\alpha}\right) = 0,\tag{6.8}$$

with $\alpha = 1, 2, \ldots, N$.

Variational methods are extremely powerful tools in science and mathematics, and they have been especially widely used in physics. There is an exceptionally elegant quantum version of the variational principle, which was developed by Richard Feynman. In this quantum method, one *sums* over all possible quantum-mechanical paths between two points in SpaceTime, which sum samples a variety of geometries (some very exotic). It has also been claimed that the Feynman sum takes into account all possible universes, thus giving application to the multiverse theory.

6.2. Paths in Euclidean Geometry

6.2.1. Straight Line in Two Dimensions. The equations (6.1), (6.2), and (6.8) can be generalized to systems other than mechanical ones. For example, consider an infinitesimal element of distance, $\mathrm{d}s$, in Cartesian coordinates, given in two dimensions by

$$\mathrm{d}s^2 = \mathrm{d}x^2 + \mathrm{d}y^2,$$

or (with $y = y_2$ at $x = x_2$, $y = y_1$ at $x = x_1$)

$$s = \int_{x_1}^{x_2} \sqrt{\mathrm{d}x^2 + \mathrm{d}y^2}$$

$$= \int_{x_1}^{x_2} \mathrm{d}x \sqrt{1 + \left(\frac{\mathrm{d}y}{\mathrm{d}x}\right)^2}. \tag{6.9}$$

Note that in Eq. (6.9) y and x play the roles of x and t, respectively, in Eq. (6.1), so that y_1 and y_2 (in Eq. (6.9)) are the same as x_A and x_B (in Eq. (6.1)).

We want to find the extremum of s, so we define the variational quantity

$$\mathcal{L}\left(y, \frac{\mathrm{d}y}{\mathrm{d}x}\right) = \sqrt{1 + \left(\frac{\mathrm{d}y}{\mathrm{d}x}\right)^2}. \tag{6.10}$$

If we use Eq. (6.2), we have $\frac{\mathrm{d}\mathcal{L}}{\mathrm{d}y} = 0$, and

$$\frac{\partial \mathcal{L}}{\partial \left(\frac{\mathrm{d}y}{\mathrm{d}x}\right)} = \frac{\left(\frac{\mathrm{d}y}{\mathrm{d}x}\right)}{\sqrt{1 + \left(\frac{\mathrm{d}y}{\mathrm{d}x}\right)^2}}.$$

In this case, x plays the role of time in Eq. (6.2), so that

$$\frac{\mathrm{d}}{\mathrm{d}x}\left[\frac{\left(\frac{\mathrm{d}y}{\mathrm{d}x}\right)}{\sqrt{1 + \left(\frac{\mathrm{d}y}{\mathrm{d}x}\right)^2}}\right] = 0, \tag{6.11}$$

or

$$\frac{\left(\frac{\mathrm{d}y}{\mathrm{d}x}\right)}{\sqrt{1 + \left(\frac{\mathrm{d}y}{\mathrm{d}x}\right)^2}} = C, \tag{6.12}$$

where C is a constant. Next, we obtain

$$\left(\frac{\mathrm{d}y}{\mathrm{d}x}\right)^2 = C^2\left[1 + \left(\frac{\mathrm{d}y}{\mathrm{d}x}\right)^2\right]$$

or

$$\left(\frac{\mathrm{d}y}{\mathrm{d}x}\right)^2 = \frac{C^2}{1 - C^2}. \tag{6.13}$$

In Eq. (6.13) we take the square root, finding that

$$\frac{\mathrm{d}y}{\mathrm{d}x} = \pm k, \tag{6.14}$$

where

$$k = \sqrt{C^2/(1 - C^2)}, \tag{6.15}$$

with $|C| < 1$ in order to make k real and finite. Then, after a simple integration in Eq. (6.14) we have

$$y = \pm kx + K, \tag{6.16}$$

where K is another constant. This is the equation of a straight line. We note too that we could adjust k and K so that $y = y_1$ at $x = x_1$ and $y = y_2$ at $x = x_2$.

One interesting point is that the above proof gives a straight line, but we don't know from the derivation whether a straight line is the shortest *or* longest distance between two points. All that we really know is that the path is an extremum. The next example will illustrate further this general problem.

6.2.2. Distance Between Two Points on a Sphere. We now perform a similar analysis between two points on a sphere. We begin by specifying the element of distance on the surface of a sphere of radius R:

$$\mathrm{d}s = R(\mathrm{d}\theta^2 + \sin^2\theta\,\mathrm{d}\phi^2)^{1/2} \tag{6.17}$$

in which we use the spherical coordinates R, θ, and ϕ. Now the analogue of (6.10) becomes

$$\mathcal{L} = (1 + \sin^2\theta\,\mathrm{d}\dot{\phi}^2)^{1/2} \tag{6.18}$$

where $\dot{\phi} = \frac{\mathrm{d}\phi}{\mathrm{d}\theta}$, with ϕ and θ taking the place of y and x, respectively; also, the path is between the two polar angles θ_1 and θ_2.

Then, the variational derivative becomes

$$\frac{\mathrm{d}}{\mathrm{d}\theta}\left[\frac{\sin^2\theta\dot{\phi}}{\left(1 + \sin^2\theta\dot{\phi}^2\right)^{1/2}}\right] = 0. \tag{6.19}$$

Next, set the quantity in brackets to an integration constant, c, and after some rearrangement this becomes:

$$\dot{\phi} = \frac{\mathrm{d}\phi}{\mathrm{d}\theta} = \frac{c/\sin^2\theta}{\left(1 - c^2 - c^2\cot^2\theta\right)^{1/2}}. \tag{6.20}$$

which, when integrated, becomes

$$\phi = \beta - \sin^{-1}(C\cot\theta). \tag{6.21}$$

where β and C are constants, with

$$C^2 = \frac{c^2}{1 - c^2}. \tag{6.22}$$

Then, rearranging the equation for ϕ, we find that

$$C\cot\theta = \sin(\beta - \phi) = \sin\beta\cos\phi - \cos\beta\sin\phi. \tag{6.23}$$

Multiplying by $R\sin\theta$, we obtain the equation:

$$Cz = x\sin\beta - y\cos\beta, \tag{6.24}$$

where we have introduced the Cartesian coordinates:

$$z = R\cos\theta;\ x = R\sin\theta\cos\phi;\ y = \sin\theta\sin\phi.$$

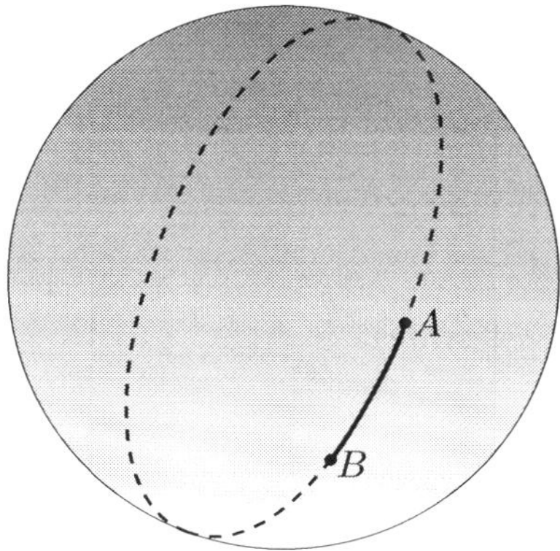

FIGURE 6.2. Great Circle on a Sphere. The shortest distance between points A and B is the solid line, while the dashed line is the longest distance between A and B.

The above equation defines a plane that intersects the origin ($x = y = z = 0$), and the intersection of this plane and the sphere gives the variational curve between the angles θ_1 and θ_2. This curve is known as the arc of a great circle, and—as shown in Fig. 6.2—it can either be the shortest or the longest distance between the two points (since both lengths satisfy the variational principle).

Most airline routes between the U.S. and Europe, across the Atlantic Ocean, follow a northern path over Newfoundland, Labrador, and occasionally Greenland. Sometimes travelers erroneously make the mistake of believing that this route is taken in order to be close to land if a flight emergency occurs. In truth, the path is taken because it follows the arc of a great circle, giving the shortest distance between the initial and final starting points for the flight. Similar considerations apply for flights across the Pacific; for example, from Los Angeles to Beijing, when the great circle path is over the Aleutian Islands and the easternmost part of Russia.

The two examples just presented are Euclidean examples of *geodesics*, a concept which comes into its own in General Relativity. For a geodesic, we must extremize the total proper time, as discussed in the next subsection. In General Relativity, a geodesic, which is a path in curved SpaceTime, is the generalization of the straight line paths of Euclidean geometry. However, the presentation in the next subsection is for *free particle* motion since no forces are assumed to be acting on this system, so we are still within the framework of SR. In General Relativity, SpaceTime is curved due to gravity which alters its metric and hence the expressions for proper time. Nevertheless, for the computation of all types of geodesics, we must extremize the proper time, which gives trajectories that are the generalizations of straight lines in Euclidean geometry. (As one book on General Relativity exclaims, "Go straight!")

6.3. Extrema in SR in Minkowski Geometry

Minkowski coordinates label the natural space in which Einstein's SR equations involving space and time are formulated. The metric for this space is given in Eq. (4.14). We assume that

The world line of a *free particle* between two *timelike* separated points *extremizes* the *proper time* between them.

In most cases, the optimum proper time is a maximum, so this is sometimes called

The Principle of Maximum Aging.

Recall Fig. (3.8) in which Bob's world line corresponds to a longer proper time than that of Alice's world line (or of any other world line between the same initial and final points).

This again is a variational problem, just like that described in section 6.1. However, now we must replace the Lagrangian by the proper time whose differential is defined in Eq. (3.18). The function is then given by

$$
\begin{aligned}
\Delta = c\tau_{AB} = c\int_A^B d\tau \\
= \int_A^B \sqrt{c^2 dt^2 - \sum_{j=1}^3 dx_j dx_j} \\
= \int_A^B \sqrt{dx_t^2 - \sum_{j=1}^3 dx_j dx_j}.
\end{aligned}
\tag{6.25}
$$

Note that we have changed notation slightly, but in an obvious way, from Eqs. (4.1) and (4.2) in chapter 4.

We can rewrite Eq. (6.25) in another way, showing an intimate connection between $d\tau$ and the Lagrangian in Eq. (6.4). We multiply the integrand of Eq. (6.25) by the constant $-mc$, which does not affect the variational derivation given by Eq. (6.2). Then, the integrand of Eq. (6.25) can be written as

$$
-mc\Delta = -mc^2 \int_A^B \sqrt{\left[1 - \left(\frac{v}{c}\right)^2\right]}\, dt,
\tag{6.26}
$$

which in the non-relativistic limit ($\left|\frac{v}{c}\right| \ll 1$) can be expressed as:

$$
-mc\Delta = \int_A^B \left[-mc^2 + \frac{1}{2}mv^2 + \ldots\right] dt.
\tag{6.27}
$$

In the bracket on the right-hand-side of Eq. (6.27), the first term is constant, so its variation must vanish, and the second term is the Newtonian kinetic energy, as in Eq. (6.3) or (6.4). Also, as previously remarked, we are extremizing the world line of a time-like free particle; thus, there is no potential energy term.

In Eq. (6.25) we then introduce a parameter σ, which is defined along the path with the limits

$$
\sigma = 0 \text{ for } \tau = \tau_A
\tag{6.28}
$$

$$
\sigma = 1 \text{ for } \tau = \tau_B,
$$

after which Eq. (6.25) becomes

$$
\begin{aligned}
\Delta = \int_A^B d\sigma \sqrt{\left(\frac{dx_t}{d\sigma}\right)^2 - \sum_{j=1}^3 \left(\frac{dx_j}{d\sigma}\right)^2} \\
= \int_0^1 d\sigma\, \mathcal{L}\left(\frac{dx_\alpha}{d\sigma}\right); \quad \alpha = t, 1, 2, 3,
\end{aligned}
\tag{6.29}
$$

where

$$\mathcal{L}\left(\frac{\mathrm{d}x_\alpha}{\mathrm{d}\sigma}\right) = \sqrt{\left(\frac{\mathrm{d}x_t}{\mathrm{d}\sigma}\right)^2 - \sum_{j=1}^{3}\left(\frac{\mathrm{d}x_j}{\mathrm{d}\sigma}\right)^2}. \tag{6.30}$$

Recall that $x_t = ct$. Note too the identity

$$\mathcal{L} = \frac{c\mathrm{d}\tau}{\mathrm{d}\sigma}. \tag{6.31}$$

The variational equation, Eq. (6.2), becomes

$$\frac{\mathrm{d}}{\mathrm{d}\sigma}\left[\frac{\partial\mathcal{L}}{\partial\left(\frac{\mathrm{d}x_\alpha}{\mathrm{d}\sigma}\right)}\right] - \frac{\partial\mathcal{L}}{\partial x_\alpha} = 0; \quad \alpha = t, 1, 2, 3, \tag{6.32}$$

where the time t in Eq. (6.2) has been replaced by the parameter σ. It is also clear that

$$\frac{\partial\mathcal{L}}{\partial x_\alpha} = 0, \tag{6.33}$$

which from Eq. (6.32) implies that

$$\frac{\mathrm{d}}{\mathrm{d}\sigma}\left[\frac{\partial\mathcal{L}}{\partial\left(\frac{\mathrm{d}x_\alpha}{\mathrm{d}\sigma}\right)}\right] = 0; \quad \alpha = t, 1, 2, 3. \tag{6.34}$$

We then have

$$\frac{\partial\mathcal{L}}{\partial\left(\frac{\mathrm{d}x_\alpha}{\mathrm{d}\sigma}\right)} = \frac{\pm\frac{\mathrm{d}x_\alpha}{\mathrm{d}\sigma}}{\sqrt{\left(\frac{\mathrm{d}x_t}{\mathrm{d}\sigma}\right)^2 - \sum_{j=1}^{3}\left(\frac{\mathrm{d}x_j}{\mathrm{d}\sigma}\right)^2}} \tag{6.35}$$

$$= \pm\frac{\left(\frac{\mathrm{d}x_\alpha}{\mathrm{d}\sigma}\right)}{\mathcal{L}},$$

where the $+$ sign is for $\alpha = t$ (time) and the $-$ sign is for $\alpha = j = 1, 2, 3$ (space). The sign doesn't matter, and we find that

$$\frac{\mathrm{d}}{\mathrm{d}\sigma}\left[\frac{\left(\frac{\mathrm{d}x_\alpha}{\mathrm{d}\sigma}\right)}{\mathcal{L}}\right] = 0; \quad \alpha = t, 1, 2, 3. \tag{6.36}$$

Now multiply Eq. (6.36) by $\frac{\mathrm{d}\sigma}{\mathrm{d}\tau}$, giving

$$\frac{\mathrm{d}}{\mathrm{d}\tau}\left[\frac{\left(\frac{\mathrm{d}x_\alpha}{\mathrm{d}\sigma}\right)}{\mathcal{L}}\right] = 0. \tag{6.37}$$

We next use Eq. (6.31), converting Eq. (6.37) to

$$\frac{\mathrm{d}}{\mathrm{d}\tau}\left[\frac{\mathrm{d}x_\alpha}{\mathrm{d}\sigma}\left(\frac{\mathrm{d}\sigma}{c\,\mathrm{d}\tau}\right)\right] = 0, \tag{6.38}$$

or

$$\frac{\mathrm{d}^2}{\mathrm{d}\tau^2}x_\alpha = 0; \quad \alpha = t, 1, 2, 3. \tag{6.39}$$

However, we also recall Eqs. (5.13) and (5.42), which imply that

$$\frac{\mathrm{d}P}{\mathrm{d}\tau} = 0, \tag{6.40}$$

which means that the four-momentum

$$P = \left(\vec{p}, \frac{E}{c} \right) \tag{6.41}$$

is conserved in *any inertial frame* for *free particle motion*. Thus we have established that

> The relativistic energy and spatial (vector) momentum are separately conserved
> in any intertial frame in SR, for a single free particle.

Moreover, for a single particle the quantity that is conserved in *transforming between reference frames* is

$$\vec{p} \cdot \vec{p} - \frac{E^2}{c^2} = -m_\circ^2 c^2, \tag{6.42}$$

which is Eq. (5.36). The generalization of these equations to the many-particle situations encountered in scattering will be presented in the next chapter.

Finally, in order to take into account interactions, we must generalize our definition of proper time. Of special interest is gravity, which takes us into General Relativity, and for which you can obtain the energy, and angular momentum again using a variational method applied to the appropriate proper time. (See problem #2 for this chapter).

6.4. Problems

(1) In Fig. 6.3, a free particle ($V(x) = 0$), with mass m, moves between point A at (x_A, t_A) and point B at (x_B, t_B). Now Newton's Law demands that a free particle travels between those points with a constant velocity which must be given by $(x_B - x_A)/(t_B - t_A) = (x_B - x_A)/T$ where $T = t_B - t_A$ is the total time. This is the full straight-line path shown in the figure. Consider another path reached in an interval of time $T/2$; namely, the dashed path to point $x = X$, also shown in the figure. For this second path, along each leg the path is straight, so the velocity is constant. Thus the velocity along the first leg is $(X - x_A)/(T/2)$ while along the second leg the velocity is $(x_B - X)/(T/2)$.

(a) Show, for the second path, that the action defined by Eqs. (6.1) and (6.3) is given by

$$S = \frac{m\left((X - x_A)^2 + (x_B - X)^2\right)}{T}.$$

(b) Show, for the path of extremal action, $\frac{\mathrm{d}S}{\mathrm{d}X} = 0$, that

$$X = \frac{(x_B + x_A)}{2}.$$

Thus, the second path must satisfy Newton's Law.

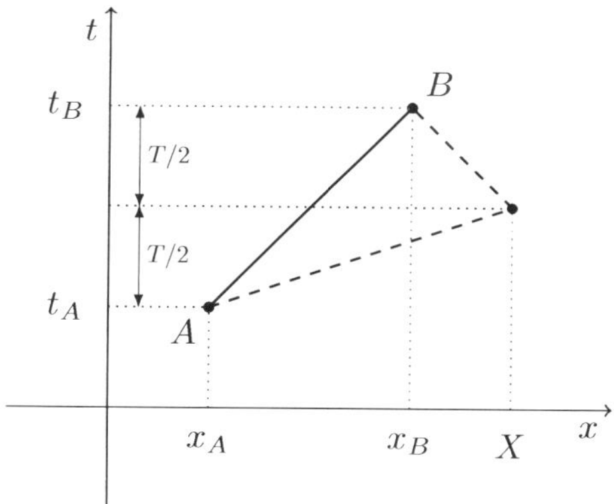

FIGURE 6.3. Demonstration of the Variational Method. The full line gives the Newtonian path from A to B. The dashed line gives an arbitrary variation in the motion from A to B. Note that the midpoint of the full line is at $X = (x_A + x_B)/2$.

(2) In General Relativity, the metric used near a Black Hole of mass M, called the Schwarzschild Metric, is given by:

$$\mathrm{d}\tau = \left[\left(1 - \frac{2M}{r}\right)\mathrm{d}t^2 + \frac{\mathrm{d}r^2}{\left(1 - \frac{2M}{r}\right)} - r^2\mathrm{d}\phi^2\right]^{1/2},$$

where r is a radial distance, t is the time, and ϕ is an azimuthal angle. (In this metric, c, the speed of light, and G, the gravitational constant, have both been set to unity.) We can then specify the Lagrangian as:

$$\mathcal{L} = \frac{d\tau}{d\sigma} = \left[\left(1 - \frac{2M}{r}\right)\left(\frac{dt}{d\sigma}\right)^2 + \frac{\left(\frac{dr}{d\sigma}\right)^2}{\left(1 - \frac{2M}{r}\right)} - r^2\left(\frac{d\phi}{d\sigma}\right)^2\right]^{1/2},$$

where, as in Eq. (6.29), σ is a parameter along the path.

(a) For the variational equation with respect to $\frac{dt}{d\sigma}$, prove that

$$\left(1 - \frac{2M}{r}\right)\left(\frac{dt}{d\tau}\right) = \text{constant}.$$

(b) Letting $r \to \infty$ in the above equation puts us into the regime of SR, where you should prove that (for $c = 1$)

$$\frac{dt}{d\tau} = \frac{E}{m}.$$

(c) Then, argue that

$$\frac{E}{m} = \left(1 - \frac{2M}{r}\right)\left(\frac{dt}{d\tau}\right) = \text{constant}.$$

(d) Next, from the variational principle for $\frac{d\phi}{d\sigma}$, prove and argue that

$$\frac{L}{m} = r^2\frac{d\phi}{d\tau} = \text{constant},$$

where L is the relativistic angular momentum (not to be confused with the Lagrangian).

Chapter **7**

The Threshold Problem and Other SR Reaction Problems in Elementary-Particle Physics

7.1. Preliminaries

For a many free-particle system, we define for the total energy and spatial momentum, respectively:

$$\mathcal{E} = E_1 + E_2 + \ldots E_N \tag{7.1}$$

$$\vec{\mathcal{P}} = \vec{p}_1 + \vec{p}_2 + \ldots \vec{p}_N, \tag{7.2}$$

where N is the total number of particles, and (from Eqs. (5.25) and (5.37))

$$E_j = \sqrt{p_j^2 c^2 + m_{\circ_j}^2 c^4} = \frac{m_{\circ_j} c^2}{(1 - v_j^2/c^2)^{1/2}} \tag{7.3}$$

and

$$\vec{p}_j = \frac{m_{\circ_j} v}{(1 - v_j^2/c^2)^{1/2}}, \tag{7.4}$$

where $m_{\circ_j}$ and v_j are, respectively, the rest mass and velocity of particle j. Also, note that by construction

$$\left(\vec{\mathcal{P}}, \frac{\mathcal{E}}{c} \right) \tag{7.5}$$

is a four vector, with the relativistic invariant

$$\mathcal{I} = \vec{\mathcal{P}} \cdot \vec{\mathcal{P}} - \frac{\mathcal{E}^2}{c^2}. \tag{7.6}$$

The quantity $\mathcal{I}$ in Eq. (7.6) is an invariant in any intertial frame, and from the last chapter both $\mathcal{E}$ and $\vec{\mathcal{P}}$ are conserved in *any* given frame. (But not from frame to frame since energy and spatial momentum get mixed up when going from frame to frame).

Moreover in what follows we will *omit* the subscript zero from the *masses*, so that any mass m_j specified is a *rest mass*.

7.2. The Threshold Problem ($N = 2$)

We begin by assuming that there are two particles in both the initial and final states ($N = 2$); at the end of this subsection, we generalize to the case in which there are still only two particles in the initial state ($N = 2$), but an arbitrary number of particles in the final state ($N = n$).

The basic problem is stated as follows:

Assume a scattering system in which two particles with rest masses m_1 and m_2 interact and produce two particles with rest masses of m_3 and m_4, with

$$m_3 + m_4 \geq m_1 + m_2. \tag{7.7}$$

The scattering can be described by the equation

$$A(m_1) + B(m_2) \rightarrow C(m_3) + D(m_4); \tag{7.8}$$

where *Cap. letter* (m_j) denotes the jth particle. We then ask the question:

What is the minimum relativistic energy (E or T) of particle A initially, in the "Lab frame", required to "just" produce particles C and D in the center-of-mass frame? The lab frame is the one in which the initial Particle B is at rest, while the center-of-mass frame is the one in which the total momentum vanishes, as in Eq. (7.9).

By "just" producing we mean that in the center-of-mass (CM) frame after the scattering the two particles are just "sitting there", each with no kinetic energy and only rest mass energy.

We also assume that the interaction is very short range, extending over a very small region of space. Thus, before and after the scattering, the particles can be considered *free particles* for which the formalism of section 7.1 applies. Moreover, we let *unprimed* variables refer to the so-called *Lab frame* and *primed* variables refer to the *CM frame*, which is defined by

$$\vec{\mathcal{P}}' = 0. \tag{7.9}$$

See Fig. 7.1 and 7.2 which show the scattering before the scattering in the Lab and CM (primed) systems, respectively. From Fig. 7.2 we have the equation (see Eq. (7.9))

$$\vec{\mathcal{P}}' = \vec{p}_1' + \vec{p}_2' = 0, \tag{7.10}$$

a relation which always defines the CM system. In Fig. 7.3 we illustrate the situation in the CM system after the scattering. Particles C and D are "just sitting there" since there is "just enough" initial energy to create, or produce, them. This behavior is why this sitation is called the *threshold* case. You should convince yourself that the configuration shown in Fig. 7.3 can only exist in the CM frame in which $\mathcal{P}' = 0$.

We now construct an *invariant* in the *CM* system *after* scattering. By definition $\vec{\mathcal{P}}' = 0$ but after scattering we have

$$\mathcal{E}' = E_3' + E_4' = m_3 c^2 + m_4 c^2 = (m_3 + m_4)c^2. \tag{7.11}$$

The invariant is given by Eq. (7.6):

$$\mathcal{I} = \vec{\mathcal{P}}' \cdot \vec{\mathcal{P}}' - \frac{\mathcal{E}'^2}{c^2}, \tag{7.12}$$

where we do *not* call this quantity $\mathcal{I}'$ since it is an invariant in *any inertial frame* (and in particular, the Lab frame). Then, from Eqs. (7.10) and (7.11), we find that for Eq. (7.12) that

$$\mathcal{I} = -(m_3 + m_4)^2 \frac{c^4}{c^2}$$
$$= -(m_3 + m_4)^2 c^2. \tag{7.13}$$

We have thus constructed an invariant in any inertial frame. However, we have also used the results of the last chapter that in the CM frame, $\vec{\mathcal{P}}'$ and $\mathcal{E}'$ are constants, which can be evaluated before and after the scattering.

We next return to the Lab (unprimed) frame *before* the scattering, Fig. 7.1, for which we obtain

$$\mathcal{E} = E_1 + E_2 = E_1 + m_2 c^2, \tag{7.14}$$

where the last term follows since particle B (with mass m_2) is just sitting there at rest. Squaring and expanding Eq. (7.14) using Eq. (7.3), we find that

$$\mathcal{E}^2 = E_1^2 + 2E_1 m_2 c^2 + m_2^2 c^4$$
$$= p_1^2 c^2 + (m_1^2 + m_2^2) c^4 + 2E_1 m_2 c^2 \tag{7.15}$$

The momentum $\mathcal{P}$ (in the x-direction) is easily obtained:

$$|\vec{\mathcal{P}}| = p_1 + p_2$$
$$= p_1 + 0 = p_1, \tag{7.16}$$

where, again because particle B is at rest, it has no momentum. Next, construct the invariant for the Lab system using Eqs. (7.15) and (7.16)

$$\mathcal{I} = \vec{\mathcal{P}} \cdot \vec{\mathcal{P}} - \frac{\mathcal{E}^2}{c^2}$$
$$= p_1^2 - \frac{p_1^2 c^2 + (m_1^2 + m_2^2) c^4 + 2E_1 m_2 c^2}{c^2}$$
$$= -(m_1^2 + m_2^2) c^2 - 2E_1 m_2. \tag{7.17}$$

Now this invariant has to be exactly equal to the invariant we calculated for the CM system in Eq (7.13), so that

$$-(m_1^2 + m_2^2) c^2 - 2E_1 m_2 = -(m_3 + m_4)^2 c^2$$

or

$$E_1 = \frac{\left[(m_3 + m_4)^2 - (m_1^2 + m_2^2)\right] c^2}{2m_2}. \tag{7.18}$$

Then, from Eq. (5.34),

$$T_1 = E_1 - m_1 c^2 = E_1 - \frac{2m_1 m_2 c^2}{2m_2}, \tag{7.19}$$

Eq. (7.19) can also be written as

$$T_1 = \frac{\left[(m_3 + m_4)^2 - (m_1 + m_2)^2\right] c^2}{2m_2}. \tag{7.20}$$

Then, from the condition in Eq. (7.7), it is clear that $T_1 \geqq 0$.

We can also generalize Eqs. (7.18) and (7.20) to the case in which there are n *particles* in the *final* CM system, with

$$m_1^{(f)}, m_2^{(f)}, m_3^{(f)}, m_4^{(f)} \ldots m_n^{(f)}, \tag{7.21}$$

$$
\begin{array}{ccc}
m_1 & v_1 & m_2 \\
\bullet\!\!\longrightarrow & - - - - - - - - & \bullet\; - - - - \to x \\
A & & B
\end{array}
$$

FIGURE 7.1. The Lab (laboratory) frame before the scattering. Note that particle B is at rest, which defines the Lab system for the threshold case.

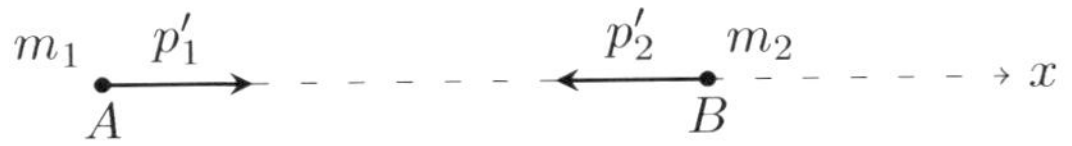

$$
\begin{array}{ccc}
m_1 \quad p_1' & & p_2' \quad m_2 \\
\bullet\!\!\longrightarrow \;\; - - - - - & \longleftarrow\!\bullet\; - - - - \to x \\
A & & B
\end{array}
$$

FIGURE 7.2. The CM (center of mass) frame before the scattering. This frame is always defined by $\vec{\mathcal{P}}' = \vec{p}_1{}' + \vec{p}_2{}' = 0$ for the two-body problem. Note: for some problems the CM frame will be the same as the Lab frame.

$$
\begin{array}{cc}
m_3 & m_4 \\
\bullet & \bullet \\
C & D
\end{array}
$$

FIGURE 7.3. The CM frame after the scattering. Both particles are at rest, each with no kinetic energy and only rest mass energy.

with

$$\sum_{i=1}^{n} m_i^{(f)} \geqq m_1 + m_2. \tag{7.22}$$

The generalizations of Eqs. (7.18) and (7.20) are

$$E_1 = \left[\left(\sum_{i=1}^{n} m_i^{(f)} \right)^2 - (m_1^2 + m_2^2) \right] c^2/2m_2 \tag{7.23}$$

and

$$T_1 = \left[\left(\sum_{i=1}^{n} m_i^{(f)} \right)^2 - (m_1 + m_2)^2 \right] c^2/2m_2 \tag{7.24}$$

Note that $n = 1$ is allowed. Also, particles #2, #3 and #4 can not be massless (e.g., photons) since Figs. 7.1 and 7.3 show particles at rest in a particular frame, which is not allowed for photons. This would completely rule out photons for the second particle and for any of the final particles. However, if particle #1 is allowed to be a photon, this can apparently cause problems with the CM frame velocity, as discussed in Appendix C.1. Nevertheless, as we show in Appendix C.2, one can reformulate the frame velocity problem and prove that particle #1 can be a photon. This important result follows since one can perform an inverse Lorentz transformation to go from the final CM frame (where no photons are allowed) to the initial Lab frame. This also agrees with the formulation presented in Ref. 8, Page 95. Problem #2 at the end of the chapter gives an example of the case in which particle #1 is a photon. For an example of photons in the final state, see subsection 7.3.1. Moreover, for all arrangements of particles, Eq (7.22) must be satisfied.

Feynman, on pages 25-4 and 25-5 of the second book of his famous lecture series[4], has an excellent practical example of the general n-particle system postulated in Eq. (7.21). Feynman was interested in how to use the acceleration of protons in the Bevatron at Berkeley to produce anti-protons, and he considered the simplest proton-proton scattering experiments which had at least one anti-proton in the final state, thus allowing several possibilities:

$$p + p \rightarrow p + \bar{p} \tag{7.25a}$$

$$p + p \rightarrow p + p + \bar{p} \tag{7.25b}$$

$$p + p \rightarrow p + p + p + \bar{p}, \tag{7.25c}$$

where p is a proton and $\bar{p}$ is an anti-proton. Since protons are baryons and anti-protons are anti-baryons, and since baryon number must be conserved in all reactions, reactions (7.25a) and (7.25b) are not allowed. Thus the only allowed reaction is (7.25c), for which Eq. (7.22) is clearly satisfied. Then, substituting the proton and anti-proton masses (which are equal) into Eqs. (7.23) and (7.24), we obtain

$$E_1 = \left[(4M)^2 - 2M^2\right] c^2/2M = 7Mc^2, \tag{7.26a}$$

or

$$T_1 = 6Mc^2. \tag{7.26b}$$

The proton mass is $M = 938$ MeV, giving $T_1 = 5.6$ GeV. (See problem #1 of this chapter for a discussion of mass units in particle physics, involving MeV and GeV.) Feynman remarks that the Bevatron accelerator at Berkeley was designed to give about 6.2 GeV of energy to the colliding protons, surely sufficient to allow for the production of antiprotons.

There is one piece of unfinished business. In our derivations we have assumed that the Lab and CM frames are inertial frames with respect to one another. For $N = 2$, this is certainly true in the classical case, for which the velocity of the CM is trivially given by

$$V_{c1} = \frac{m_1 v_1}{(m_1 + m_2)}, \tag{7.27}$$

where m_1, m_2 and v_1 are all constants. We need to establish the generalization of Eq. (7.27) to the relativistic case, which we accomplish in Appendix C.1. However, as mentioned previously, in C.1 there is a problem if particle #1 is a photon, a difficulty that we resolve in Appendix C.2.

7.3. Decay and Scattering Problems in Elementary Particle Physics

In the examples that follow, the CM scattering system is the same as the Lab system.

7.3.1. Decay of a Single Particle A into Two Photons $\gamma + \gamma$.

Consider the decay

$$A \rightarrow \gamma + \gamma, \tag{7.28}$$

where, e.g., A could be a π° meson. Note that A is at rest. In the CM system, which is the same as the Lab system, *before* and after the decay, we have

$$\vec{p}_A = \vec{p}_{\gamma_1} + \vec{p}_{\gamma_2} = 0, \tag{7.29}$$

with

$$E_{\gamma_1} = |\vec{p}_{\gamma_1}|c = E_{\gamma_2} = |\vec{p}_{\gamma_2}|c = pc = E_\gamma \tag{7.30}$$

and

$$E_A = E_{\gamma_1} + E_{\gamma_2}, \tag{7.31}$$

or
$$m_A c^2 = 2pc = 2E_\gamma. \tag{7.32}$$

However, we know from Quantum Mechanics that we can also express the photon's energy as
$$E = h\nu, \tag{7.33}$$

where h is Planck's constant and ν is the frequency of the photon. Thus we have
$$m_A c^2 = 2h\nu, \tag{7.34}$$

or
$$\nu = \frac{m_A c^2}{2h}. \tag{7.35}$$

Note that each photon in this decay has the same frequency since by conservation of momentum each has the same magnitude for momentum, and from $E = pc$ their energies and, hence, frequencies are identical.

7.3.2. Decay of a Particle A into Two Identical Particles $(B + B)$.

We have $(m_{B_1} = m_{B_2} = m_B)$
$$A \rightarrow B + B, \tag{7.36}$$

where in the CM system
$$\vec{p}_{B_1} = -\vec{p}_{B_2}; \quad (|\vec{p}_{B_1}| = |\vec{p}_{B_2}| = p_B). \tag{7.37}$$

Balancing energies before and after the decay
$$m_A c^2 = \sqrt{m_{B_1}^2 c^4 + p_{B_1}^2 c^2} + \sqrt{m_{B_2}^2 c^4 + p_{B_2}^2 c^2}$$
$$= 2\sqrt{m_B^2 c^4 + p_B^2 c^2}, \tag{7.38}$$

or
$$p_B = |\vec{p}_B| = \frac{c}{2}\sqrt{(m_A^2 - 4m_B^2)}. \tag{7.39}$$

If the B particles are both photons, $m_B = 0$, and
$$p_B = \frac{cm_A}{2}, \tag{7.40}$$

which agrees with Eq. (7.32) in the previous subsection.

7.3.3. Minimum Energy of Two Colliding Particles of Equal Mass, Producing a Very Heavy Particle.

For example, consider the electron-positron annihilation to produce the $Z_\circ$ boson and other particles:
$$e^+ + e^- \rightarrow Z^\circ + \text{(other particles)} \tag{7.41}$$

In this case we begin in the CM frame (which can also be considered the Lab frame) in which we have
$$\vec{p}_{e^+} + \vec{p}_{e^-} = 0, \tag{7.42}$$

so that
$$|\vec{p}_{e^+}| = |\vec{p}_{e^-}| = p_e. \tag{7.43}$$

Energy conservation then gives

$$2E_e = 2\sqrt{m_e^2 c^4 + p_e^2 c^2} = E_{Z_\circ} + E_\chi \tag{7.44}$$

where E_χ is the energy of the other particles. Thus we obtain

$$E_e \geqq \frac{1}{2} M_{Z_\circ} c^2. \tag{7.45}$$

The Z° boson has been determined to have a mass of ~ 91 GeV $= 91 \times 10^3$ MeV (or MeV/c^2), or

$$E_e \gtrsim 45,500 \text{ MeV}, \tag{7.46}$$

which is an enormous acceleration energy. In the actual experiment at the CERN collider in Geneva, the particles used were a proton and an anti-proton, with $E_P = 270$ GeV, giving a head-on collision energy of 540 GeV. This is about six times the energy required, but the extra energy is necessary since only a fraction of it goes into the quark/anti-quark annihilation in the proton/anti-proton collision.

The details of the Z° detection make fascinating reading. The experiment was headed up by the Italian Carlo Rubbia, who had a large group of physicists, graduate students, and post-docs. However, it was necessary to find a way to store the anti-protons before collision so that they would not annihilate with protons from ordinary materials. A method known as Stochastic Cooling accomplished this task, and it was developed by the brilliant Dutch engineer Simon van der Meer. Without Stochastic Cooling, the experiment could not have been done (or would have been tremendously more difficult). Thus, when the Nobel Prize for Physics was awarded in 1984 for the discovery of the Z° (along with the $W^\pm$ particles), Rubbia and van der Meer *shared* the prize. This was a remarkable development since it was the first time that a technical person, not a physicist, shared the Physics Nobel Prize for a fundamental discovery.

Finally, we should say a word about the CM system being the Lab systen in these last three examples. In the threshold case, we considered bombarding a particle on another particle at rest, and we referred to this arrangement as the Lab system since it is the natural arrangement one would encounter in a laboratory experiment. However, in the last three examples, the natural laboratory situation is the CM system, as we see in the Decay Cases 7.3.1 and 7.3.2, or the collision of two equal mass particles in 7.3.3. Other, less symmetrical examples can also be analyzed in which cases some of the formulas of section 7.1 can be used.

7.4. Problems

(1) A π^- meson (rest mass, $m = .140$ GeV$/c^2$) collides with a proton ($m = .938$ GeV$/c^2$) at rest. Find the minimum (threshold) engery required to produce a $K_\circ$ particle ($m = .498$ GeV$/c^2$) and a $\Lambda_\circ$ particle ($m = 1.115$ GeV$/c^2$). (Reaction: $\pi^- + p \to K_\circ + \Lambda_\circ$.) Remarks: You need not understand the definitions of these particle; for the solution to this problem you just need to know their masses. Recall that the threshold energy refers to the minimum total energy (or kinetic energy) of the π^- meson in the laboratory frame. In the threshold formulation in the text, the π^- meson corresponds to particle #1 and the proton to particle #2. The $K_\circ$ and $\Lambda_\circ$ particles correspond to particles #3 and #4. All masses given are rest masses. You may express your final answer as E_1 or T_1 (or both).

Also, it is usually convenient with elementary particles to express their rest masses in terms of an energy unit divided by c^2. (Thus, we shall sometimes simply say, e.g., that the rest mass of the π^- meson is .140 GeV.) Note too that $1eV$ is the energy acquired by *one electron* after being accelerated through a potential (or *voltage difference*) of 1 volt. Then, 1 MeV $= 10^6$ eV and 1 GeV $= 10^3$ MeV $= 10^9$ eV.

(2) A reaction for *photoreducing pions* is: $\gamma + p \to n + \pi^+$, where γ is a photon, p is a proton, n is a neutron, and π^+ is a positively charged pi-meson. We can use the threshold method for this case if we realize that the validity of Eq. (7.18) depends only on defining a CM frame for the final particles and a Lab frame for the initial particles. Thus, using Eq. (7.18), find the minimum energy that a photon would require in order to produce a pion in the frame in which the proton is at rest. In this threshold problem, particle #1 (the projectile) is the photon, which has zero rest mass. This is the *standard threshold problem* in which the neutron and pi-meson are at rest in the center-of-mass system after the scattering, while the proton is at rest in the Lab system before the scattering. The proton and neutron are assumed to have the same mass, namely 938 MeV, while the π^+ meson has a rest mass of 140 MeV.

(3) Consider the π-meson decays:

$$\pi^+ \to \mu^+ + \nu_\mu$$

$$\pi^- \to \mu^- + \bar{\nu}_\mu,$$

where $\pi^\pm$ are the charged π mesons, $\mu^\pm$ are the charged muons, ν_μ is a muon neutrino, and $\bar{\nu}_\mu$ is a muon anti-neutrino. You can use the first part of Eq. (7.38) (before the restriction that the masses are equal is made). Neglect the masses of the neutrino and anti-neutrino, and assume that the companion charged particles have equal masses. Work in the CM system of the decaying pions, and prove that

$$T_\mu = \frac{(m_\pi - m_\mu)^2}{2m_\pi} c^2,$$

where T_μ is the relativistic kinetic energy of the muon.

(4) *The Compton Effect.* Fig. 7.4 shows a photon scattering off of an electron in the laboratory system. The left-hand picture shows the situation just prior to scattering (with the electron

66

initially at rest), while the right hand picture shows the geometry after scattering. After scattering, the electron has a velocity $\vec{v}$, whose direction is at an angle ϕ with respect to the x-axis. The frequency of the photon is ν before scattering and ν' after scattering. Remember that for a photon (from Quantum Mechanics and SR) $E = h\nu$ and $p \equiv |\vec{p}| = E/c$, while for any particle the four-momentum is given by:

$$p_\mu : \left(\vec{p}, \frac{E}{c}\right) = \left(p_x, p_y, p_z; \frac{E}{c}\right).$$

(a) Before scattering, find the four-momentum of each particle and the total laboratory four-momentum.

(b) After scattering, find the four-momentum of each particle and the total laboratory four-momentum.

(c) By conserving four-momentum in the laboratory system you obtain three equations. (There is no z-component of four-momentum.) What are these three equations?

(d) From these three equations, prove that

$$\nu' = \frac{\nu}{\left\{1 + \frac{h\nu}{m_\circ c^2}[1 - \cos(\theta)]\right\}},$$

where $m_\circ$ is the rest mass of the electron and θ is the angle of scattering of the photon after the scattering.

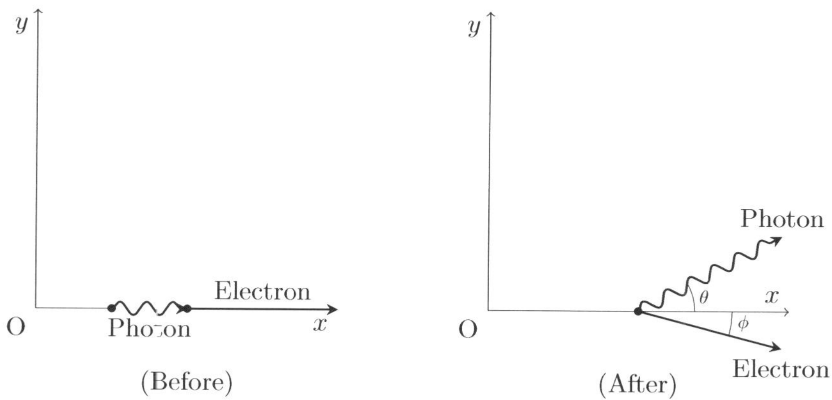

FIGURE 7.4. Geometry of the Compton effect.

(5) (a) Can a single particle decay into an single photon? Why or why not?

(b) Can a system of particles (e.g., the positronium system in which an electron and positron are briefly bound by their mutual electromagnetic attraction) decay into a single photon? Why or why not?

Chapter **8**

The Relativistic Doppler Shift

8.1. The Classical Doppler Effect

We are all aware of the Doppler shift (or effect) in classical physics. Consider the pitch (frequency) of a whistle from a train. If the train is approaching you, the pitch increases, while if the train is receding from you, the pitch decreases.

In classical physics, expressions have been derived for the Doppler effect. For a moving source approaching a stationary observer, it is found that

$$\nu_o = \frac{\nu_s}{1 - \frac{u}{c}}, \tag{8.1}$$

where ν_s is the frequency emitted by the source, ν_o is the frequency experienced by the stationary observer, $u > 0$ is the velocity of positive approach of the source, and c is the velocity of the wave being emitted (e.g.: sound or light).

Next, consider a moving observer approaching a stationary source, for which it is found that

$$\nu_o = \nu_s \left(1 + \frac{u}{c}\right), \tag{8.2}$$

where $u > 0$ is the velocity of positive approach.

In Eqs. (8.1) and (8.2), for $u > 0$, the frequency (or pitch) experienced by the observer is greater than the emitted frequency of the source ($\nu_o > \nu_s$). On the other hand, if $u < 0$ (velocity of recession), the frequency experienced by the observer is less than the emitted frequency ($\nu_o < \nu_s$). For very small u ($|\frac{u}{c}| \ll 1$), the two expressions are approximately equal, with

$$\nu_o = \nu_s \left(1 + \frac{u}{c}\right). \tag{8.3}$$

Eqs. (8.1) and (8.2) satisfy our intuition about the change in pitch. However, they do not satisfy our understanding of relativity since clearly we should get the same expression for a moving source approaching a stationary observer as for a moving object approaching a stationary source. We want to correct this weakness in classical physics by performing a proper derivation using SR.

Moreover, in classical physics one can write down an expression for motion of both the observer and source, for which

$$\nu_o = \frac{\nu_s \left(1 + \frac{u_o}{c}\right)}{\left(1 - \frac{u_s}{c}\right)}, \tag{8.4}$$

where u_o is the velocity of the observer and u_s is the velocity of the source.

8.2. The Doppler Effect in SR

In this derivation we will also incorporate a feature of Quantum Mechnics whereby, due to the wave-particle duality, the photon is the particle representation of light, for which we can specify its frequency using Eq. (7.33),

$$E = h\nu, \tag{8.5}$$

where h is Planck's constant. Then, from the photon energy-momentum expression for SR, Eq. (5.41), we have

$$p \equiv |\vec{p}| = \frac{E}{c} = \frac{h\nu}{c}. \tag{8.6}$$

Also, since the photon is a particle we write its energy-momentum four vector as

$$P = \left(\vec{p}, \frac{E}{c} \right) = \left(p_x, p_y, p_z; \frac{E}{c} \right). \tag{8.7}$$

Since we will only be concerned about a photon moving in the x-direction (as in Fig. 8.1), we obtain from Eqs. (8.6) and (8.7)

$$P = \frac{h\nu}{c}(1, 0, 0; 1), \tag{8.8}$$

where $P_x = |\vec{p}| = p = \frac{h\nu}{c}$. Next we can use the Lorentz transformation of four vectors, Eq. (4.5), to transform the momentum and energy from the moving source system (S') to the stationary observer system (S), or vice versa. (See Fig. 8.1.)

From the Lorentz transformation, Eq. (4.5d), we obtain

$$\frac{E'}{c} = \gamma \left(\frac{E}{c} - \frac{u}{c} p \right) \tag{8.9}$$

$$= \gamma \frac{h\nu}{c} \left(1 - \frac{u}{c} \right),$$

where

$$\gamma = \left(1 - \frac{u^2}{c^2} \right)^{-1/2}. \tag{8.10}$$

Then, letting $E' = h\nu'$ and $\nu' = \nu_s, \nu = \nu_o$, we have (after cancelling h from both sides of Eq. (8.9))

$$\nu_s = \nu_o \gamma \left(1 - \frac{u}{c} \right) \tag{8.11}$$

$$= \sqrt{\frac{(1 - u/c)}{(1 + u/c)}} \, \nu_o,$$

or

$$\nu_o = \sqrt{\frac{(1 + u/c)}{(1 - u/c)}} \, \nu_s. \tag{8.12}$$

Notice that for $u > 0$, $\nu_o > \nu_s$ (called a blue shift for light). Also, we get the same result if we use the Lorentz equation (4.5a). Moreover, we recover the classical result, Eq. (8.3), if we take the limit of small velocity ($|\frac{u}{c}| \ll 1$) in Eq. (8.12).

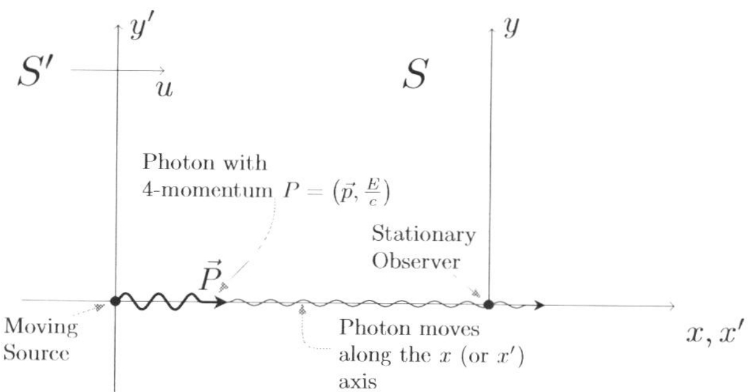

FIGURE 8.1. The Relativistic Doppler Effect. The direction of the photon is in the positive x-direction in both frames. Frame S' approaches S with a velocity u (in the x-direction), and S' lies to the left of S.

Next, consider Fig. 8.2 which is the same as Fig. 8.1, but with the direction of u reversed. For $u > 0$, positive toward the left, we obtain

$$\nu_\circ = \sqrt{\frac{(1 - u/c)}{(1 + u/c)}} \, \nu_s. \tag{8.13}$$

Thus, for $u > 0$ (a positive velocity of recession) $\nu_\circ < \nu_s$, which for light is called a red shift.

You can also put S' on the other side of S, as shown in Figs. 8.3 and 8.4. From Problems 1 and 2, show that you reproduce the Doppler shift formulas (8.13) and (8.12), respectively.

Therefore, there is in reality only one formula

$$\nu_\circ = \sqrt{\frac{(1 + u/c)}{(1 - u/c)}} \, \nu_s, \tag{8.14}$$

where $u > 0$ for a source moving *toward* a stationary observer and $u < 0$ for a source moving *away* from a stationary observer. Eq. (8.14) is also the expression one obtains for the Doppler shift of an observer moving with velocity u toward a stationary source (and $u < 0$ for an observer moving away from a stationary source). We shall now prove this for one special case, illustrated in Fig. 8.5, for

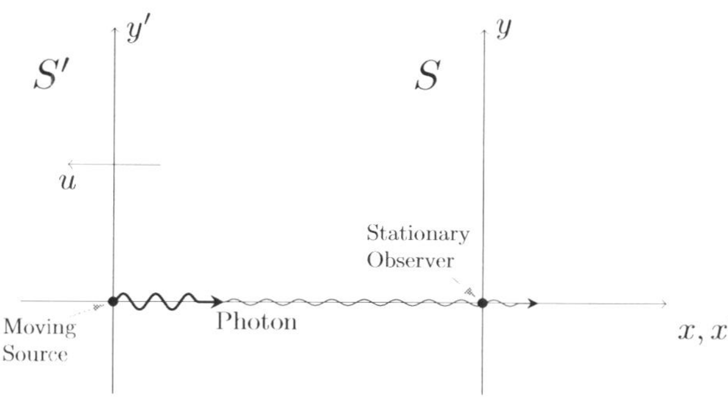

FIGURE 8.2. Fig. 8.1, with the sign of u reversed. Note that the photon moves in the positive x, or x' direction, in both frames.

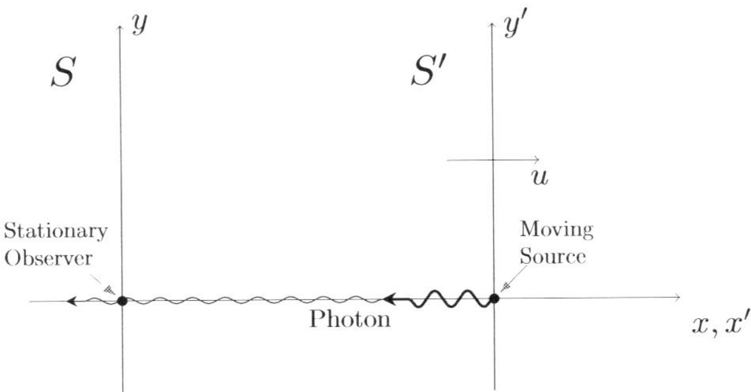

FIGURE 8.3. Same as Fig. 8.1, but with S' on the right side of S.

which we have for the inverse transformation of Eq. (4.5d)

$$\frac{E}{c} = \gamma\left(\frac{E'}{c} + \frac{u}{c}p'_x\right)$$

$$= \gamma\frac{E'}{c}\left(1 - \frac{u}{c}\right) \tag{8.15}$$

since $p'_x(\text{photon}) = -|\vec{p'}| = -E'/c$. Then, using Eq. (8.6) we obtain

$$\nu = \gamma\nu'\left(1 - \frac{u}{c}\right)$$

$$= \nu'\sqrt{\frac{(1 - u/c)}{(1 + u/c)}}, \tag{8.16}$$

or

$$\nu_\circ = \sqrt{\frac{(1 + u/c)}{(1 - u/c)}}\,\nu_s, \tag{8.17}$$

which is the same as Eq. (8.14). Other cases for a moving observer can be constructed, but one obtains the same result. Thus, there is *one expression* for the Doppler shift, Eq. (8.17), with u the *relative velocity* between the source and the observer, which is positive for a source or observer moving towards one another; negative for a recession. *This is indeed relativity!*

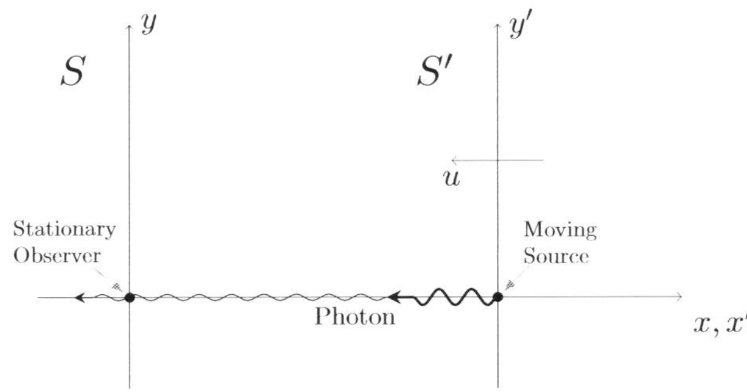

FIGURE 8.4. Same as Fig. 8.2, but with S' on the right side of S

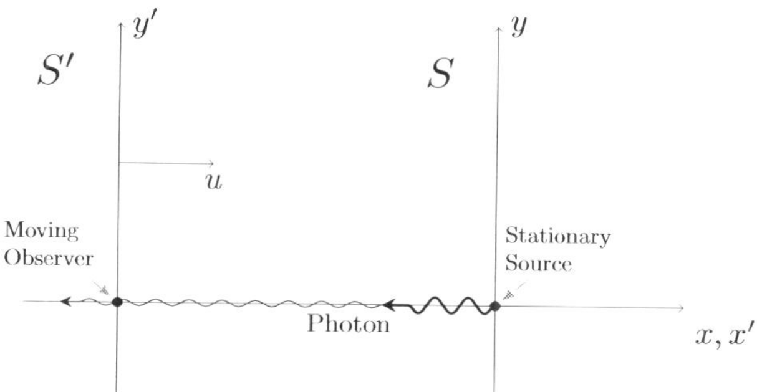

FIGURE 8.5. Example of a moving observer and a stationary source

We now make a remark about the terms "blue shift" and "red shift", which have been used for more than a century to describe spectral frequencies measured in astronomy. A blue shift simply means that the frequency experienced by the observer is greater than the emission frequency of the source. The name originated since the blue end of the visual (white) spectrum has a higher frequency than the center of the spectrum. Similarly, a red shift means that the frequency experienced by the observer is less than that emitted by the source. Again, the name originated since the red end of the visible spectrum has a smaller frequency than that of the center of the spectrum.

Also, the knowledgable reader may wonder why we have focused on one ray of light, which travels from S' to S, rather than on an envelope of light waves, with rays travelling in all possible directions. The answer, of course, is that in Quantum Mechanics, light is both a particle and a wave (the wave-particle duality), and which aspect you focus on depends upon your experiment and your purpose. Here we choose to emphasize the particle aspect of light, whose manifestation is the photon. On the other hand, in Appendix A, in deriving the Lorentz equation, it was more convenient to focus on the wave nature of light; hence, there we considered the envelope of light wave emanating from a source at the origin of a coordinate system. Moreover, the objection to focusing on photon trajectories could also be leveled, in previous chapters, at various particles which also have both wave-like and particle-like properties (and, hence, are part of the wave-particle duality or more generally the Principal of Complimentarity).

In Appendix D, we treat the general case of a source separated from a stationary observer, at an angle, as illustrated in Figs. 8.6 and 8.7. We obtain the results

$$\nu_\circ = \nu_s \left[\gamma \left(1 - \frac{u}{c}\cos\alpha\right)\right]^{-1}, \tag{8.18}$$

and

$$\cos\alpha' = \frac{\cos\alpha - \frac{u}{c}}{1 - \frac{u}{c}\cos\alpha}, \tag{8.19}$$

where α and α' are the angles of the photons in S and S' respectively. Notice that if $\alpha = \alpha' = 0$, we recover Eq. (8.17).

Moreover, another special case arises for $\alpha = \pm\frac{\pi}{2}$ $\left[\alpha' = \cos^{-1}\left(\frac{-u}{c}\right)\right]$; namely,

$$\nu_\circ = \frac{\nu_s}{\gamma} = \sqrt{\left(1 - \frac{u^2}{c^2}\right)}\,\nu_s, \tag{8.20}$$

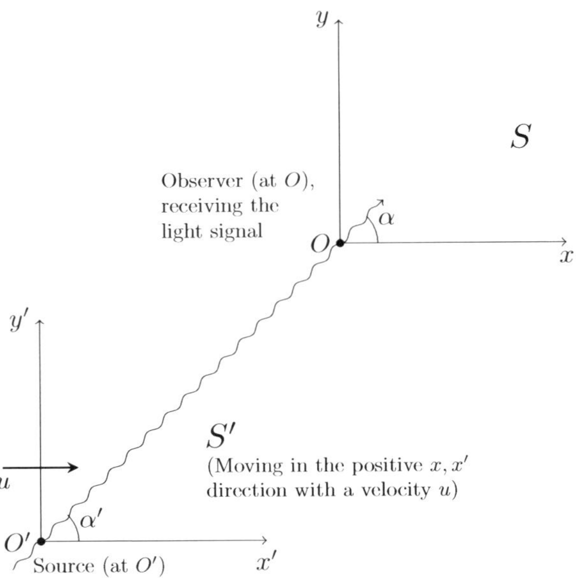

FIGURE 8.6. Doppler effect, with the moving source and stationary observer separated by an angle, where α and α' pertain to the angles of the photon in S and S', respectively. The two-dimensional picture depicted can always be obtained by a translation and rotation of axes.

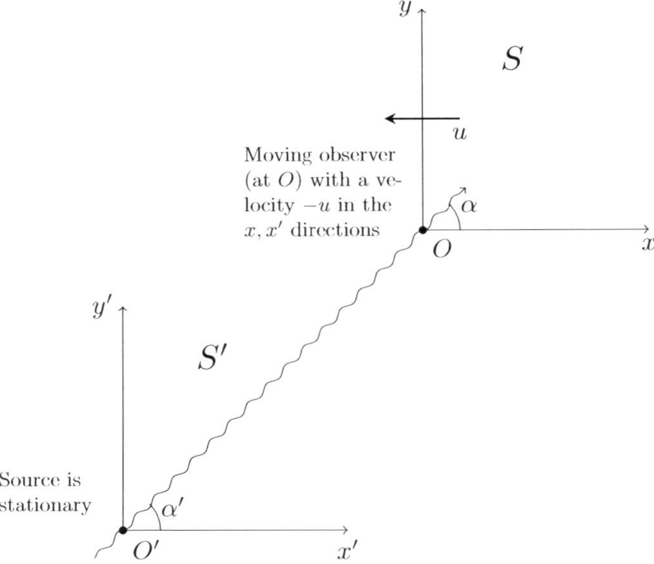

FIGURE 8.7. Revision of Fig. 8.6 in which the source is stationary and the observer moves with a velocity $-u$ in the x, x' directions.

which is called the *Transverse Doppler Effect*. It is a purely relativistic shift, which has no classical analogue. (To first-order in $|\frac{u}{c}|$, $\nu_\circ = \nu_s$,) Also, the change in frequency, when you view a source from a perpendicular direction, is always a *red shift*, regardless of the sign of u.

8.3. The Red-Shift Parameter, z

The Doppler Effect has enormous applications, in everything from atomic and nuclear spectroscopy to medicine. But perhaps its most interesting and important applications are in the fields of Astronomy, Astrophysics, and Cosmology. This arises because in the early part of the twentieth century Vestro Slipher and Edward Hubble observed that almost all galaxies recede from us and our Milky Way galaxy, giving rise to a red shift in frequencies. One of the very few exceptions is Andromeda, the nearest large galaxy, which due to the mutual gravitational attractions of the Milky Way and Andromeda, is moving towards us, giving a blue shift in frequencies. However, the fact that most galaxies tend to recede from one another is one of the main pillars of the Big Bang cosmology. We will return to Hubble shortly, but first some important preliminaries.

For light (or indeed any wave motion) we have

$$\lambda\nu = c, \tag{8.21}$$

where ν is the frequency of the wave, λ is its wave length, and c is its speed. Notice that the units of Eq. (8.21) are

$$\text{length} \cdot \frac{1}{\text{time}} = \frac{\text{length}}{\text{time}} = \text{velocity} \tag{8.22}$$

From Eq. (8.21), frequency and wave length are inversely proportional to one another, so—say for a red shift—the change in frequency decreases and the change in wave length increases; i.e.,

$$\text{if } \Delta\nu < 0, \ \Delta\lambda > 0, \tag{8.23}$$

and vice versa.

It is convenient to define a red-shift parameter, z, as follows:

$$\text{Red-Shift Parameter} = z \equiv \frac{\lambda_\circ - \lambda_s}{\lambda_s} = \frac{\Delta\lambda}{\lambda}, \tag{8.24}$$

where λ_s is the wave length of the source, $\lambda_\circ$ is the wave length experienced by the observer, $\Delta\lambda$ is the change in wave length,

$$\Delta\lambda \equiv \lambda_\circ - \lambda_s, \tag{8.25}$$

and $\lambda \approx \lambda_s$ or $\lambda_\circ$ (if the difference is assumed to be small). Then, from Eq. (8.21), Eq. (8.24) can be reexpressed as

$$z = \frac{\frac{c}{\nu_\circ} - \frac{c}{\nu_s}}{\frac{c}{\nu_s}} = \frac{\nu_s(\nu_s - \nu_\circ)}{\nu_\circ \nu_s},$$

or

$$z = \frac{\nu_s - \nu_\circ}{\nu_\circ} = \frac{-\Delta\nu}{\nu_\circ} = -\frac{\Delta\nu}{\nu}, \tag{8.26}$$

where

$$\Delta\nu \equiv \nu_\circ - \nu_s \tag{8.27}$$

In differential form, using Eq. (8.21), we find that

$$d\lambda = -\frac{c}{\nu^2}d\nu$$

$$\frac{d\lambda}{\lambda} = -\frac{c}{\nu^2}\frac{d\nu}{(c/\nu)},$$

or

$$\frac{d\lambda}{\lambda} = -\frac{d\nu}{\nu}, \tag{8.28}$$

which can be seen from Eqs. (8.24) and (8.26).

Then, from our Doppler shift formula, Eq. (8.17), we obtain from Eq. (8.26)

$$z = \frac{\nu_s}{\nu_\circ} - 1 = \sqrt{\frac{1 + \frac{u}{c}}{1 - \frac{u}{c}}} - 1, \tag{8.29}$$

where $u > 0$ now means a positive velocity of recession, which applies to almost all galaxies*. Eq. (8.29) is a general formula but it is often convenient to work with small velocities of recession, so that—using only the first term of the Taylor expansion, we have

$$z \approx \left(1 + \frac{1}{2}\frac{u}{c}\right)\left(1 + \frac{1}{2}\frac{u}{c}\right) - 1,$$

or

$$z \approx \frac{u}{c} \qquad \left(\frac{u}{c} \ll 1\right). \tag{8.30}$$

which we identify as the classical result. (See Eq. (8.3).) Now either Eq. (8.29) or (8.30) can be used, but we emphasize again that Eq. (8.30) is only valid for small velocities of recession.

Moreover, from Eqs. (8.24) and (8.26)

$$\text{if } z < 0, u < 0, \Delta\lambda < 0, \text{ and } \Delta\nu > 0 \text{ (Blue Shift)}, \tag{8.31a}$$

and

$$\text{if } z > 0, u > 0, \Delta\lambda > 0, \text{ and } \Delta\nu < 0 \text{ (Red Shift)}. \tag{8.31b}$$

Also, from Eq. (8.29), we have

$$z + 1 = \sqrt{\frac{1 + \frac{u}{c}}{1 - \frac{u}{c}}}$$

and

$$(z + 1)^2 = \frac{1 + \frac{u}{c}}{1 - \frac{u}{c}}$$

or

$$(z + 1)^2 \left(1 - \frac{u}{c}\right) = \left(1 + \frac{u}{c}\right),$$

giving

$$\frac{u}{c} = \frac{(z + 1)^2 - 1}{(z + 1)^2 + 1}. \tag{8.32}$$

For z small ($|z| \ll 1$)

$$\frac{u}{c} = \frac{z^2 + 2z}{z^2 + 2z + 2}$$

$$\approx \frac{2z}{2} = z,$$

which is Eq. (8.30).

The shift in frequency (or wavelength) can be determined using a spectrometer, which gives the value of z, which—in turn—gives the value of the recession velocity. Thus, the measurment of z is

---

*Note that the sign convention on u in Eq. (8.29) is reversed from the sign convention used in Eq. (8.17).

completely equivalent to the determination of the recession velocity u. In astronomy, graphs for the variable u are often given using z, the red-shift parameter.

8.4. Hubble's Law and the Hubble Distance

By analyzing, using a spectrometer, the change in frequency[†] of light from certain elements or compounds in the stars of a distant galaxy, astronomers can measure the (recession) speed of that galaxy. Thus, obtaining the velociy of recession of a galaxy is a straight-forward procedure employing basic spectroscopy and the Doppler shift.

However, astronomers have also long been interested in measuring the distance to remote galaxies, and this has proven to be a much more formidable undertaking. There is a long history of this activity, which we will not detail here. Suffice it to say, the main method involves using so-called "standard candles". The idea is first to measure the distance to a near-by object using some well-tested method. Then, you find some other characteristic of the object, which characteristic also gives the correct distance. The next phase of the process involves looking for the second characteristic in an object much further away than the first object, which allows the distance to the second object to be obtained. For example, the early measurements of many stars in the Milky Way galaxy and Small Magellanic Cloud[‡] were determined using a variation of a method called parallax. Then, in 1912 it was discovered by Henrietta Leavitt that certain stars in the Small Magellanic Cloud exhibited variations in brightness, which period is of the order of days. These stars are called Cepheids, and Leavitt discovered that their periods were related in a linear fashion to their intrinsic brightness. Next, when a Cepheid was discovered in a galaxy further away, it was possible from its period to obtain its intrinsic brightness. Finally, by comparing its intrinsic brightness to its apparent brightness (using a $1/r^2$ formulation), it was possible to determine the distance of the galaxy in which the second, more distant Cepheid resides. Thus, in principal, we construct a ladder of standard candles, which, step-by-step, allows you to measure the distances to far-away galaxies.

It should be mentioned too that, in 1952 at a major astronomy meeting, Walter Baade announced that he had found an error in the previous distance calculations of Cepheid variable stars. This discovery, which received considerable astonishment at the meeting, arose because there were two types of Cepheids and resulted in a doubling of the distance to the Andromeda galaxy, and also a doubling of both the size of the known Universe and its age. (See Eq. (8.35).) This example was important since it illustrates the effect of possible errors on the major rungs of the standard candles.

In 1929, Edwin Hubble announced the famous result that the speed of recession of a galaxy, v, is (roughly) proportional to its distance from us, d:

$$v = Hd, \tag{8.33}$$

where H is the Hubble constant, whose present value $H_\circ$ is

$$H_\circ = (70 \pm 7)\frac{\text{km}}{\text{sec}}\,\frac{1}{\text{MPc}}; \tag{8.34}$$

and MPc = MegaParsec = 10^6 Parsec; 1 Parsec = 3.26 light years, a convenient unit of length used in astronomy. Thus, as the distance of a galaxy increases its velocity also increases. Eq. (8.33) has at least two important consequences. First, if we know the velocity of recession of a galaxy (or equivalently, its red shift z) we can estimate its distance and vice verse. Second, even though H

[†]In practice, often the data analyzed are absorption spectra (rather than emission spectra) since the atmosphere of the stars absorb the light emitted internally.

[‡]a "satellite" sub-galaxy of the Milky Way.

is expressed in odd units, it has the dimension of $(\text{time})^{-1}$. Thus, cosmologists have been able to roughly calculate the age of the Universe using

$$\text{Present age of the Universe} = (H_\circ)^{-1} \tag{8.35}$$

$$= 14.0 \pm 1.4 \text{ Gyr.}$$

You see too that, since v increases with d, v increases without limit and eventually will equal or exceed c. The distance at which equality occurs is called the *Hubble Distance*, given by

$$d_H = \frac{c}{H_\circ}. \tag{8.36}$$

Question: does this behavior not violate one of the main tenets of SR; namely, that nothing can exceed the speed of light? The answer, surprisingly, is "No, it does not violate SR." First, in SR *the motion of objects is assumed to be in a static space.* Second, in the context of *General Relativity*, there is no objection to having two points moving away from one another at a *superluminal speed (i.e.: $v > c$) due to the expansion of space.* Then, the Hubble distance in Eq. (8.36) defines the size of our *observable universe.* The current value of the Hubble distance is 4300 ± 400 MPc. We can get no information about portions of the Universe which lie beyond the Hubble distance. See the discussions in [46], pages 16 and 39.

In Chapter 10, we give some additional examples in which superluminal velocities appear to be observed. (In one case, the superluminal motion is an optical illusion.)

8.5. Relativistic Beaming

We conclude this chapter with another application of the Doppler shift. It is well known that a uniformly radiating body moving toward you is brighter than if it is moving away from you, a phenomenon known as relativistic beaming. The Doppler shift implies that, for such a body, the energy of the photons in the forward direction is much greater than those in the backward direction, and it can be shown that the intensity of the radiation is concentrated along the direction of motion. See Fig. 8.8.

In Eq. (8.19) there are restrictions, on the angles, which become especially apparent as the velocity of the source, u, approaches the speed of light We can solve Eq. (8.19) for $\cos\alpha$, obtaining

$$\cos\alpha = \frac{\left(\frac{u}{c} + \cos\alpha'\right)}{\left(1 + \frac{u}{c}\cos\alpha'\right)}, \tag{8.37}$$

a result that was derived another way in Eq. (i) of Problem #3 for this chapter. Next, in Eq. (8.37), take the limit as $u \to c$, giving

$$\cos\alpha \to \frac{1 + \cos\alpha'}{1 + \cos\alpha'} = 1, \tag{8.38}$$

or

$$\alpha \to 0. \tag{8.39}$$

Thus, the cone of α is highly concentrated about the forward direction, regardless of the values of α' (even though as $\alpha \to 0, \alpha' \to 0$). Then, from Eq. (8.18) we have

$$\nu_s = \nu_\circ\gamma\left(1 - \frac{u}{c}\cos\alpha\right), \tag{8.40}$$

which, for $\alpha \approx 0$ and $u \approx c$, becomes

$$\nu_s \approx \nu_\circ \frac{\left(1 - \frac{u}{c}\right)}{\sqrt{1 - \frac{u^2}{c^2}}}$$

$$= \nu_\circ \sqrt{\frac{\left(1 - \frac{u}{c}\right)}{\left(1 + \frac{u}{c}\right)}}, \tag{8.41}$$

or

$$\nu_\circ \approx \nu_s \sqrt{\frac{\left(1 + \frac{u}{c}\right)}{\left(1 - \frac{u}{c}\right)}} \gg \nu_s \quad \left(\text{for } \frac{u}{c} \approx 1\right) \tag{8.42}$$

Thus, for $u \approx c$, relative to the observer (and recall too that $u > 0$ corresponds in Fig. 8.6 to a source moving toward the observer) the frequency is very large, compared to the frequency of the source. Moreover, it is highly concentrated in a very small cone in the forward direction, as we shall now demonstrate.

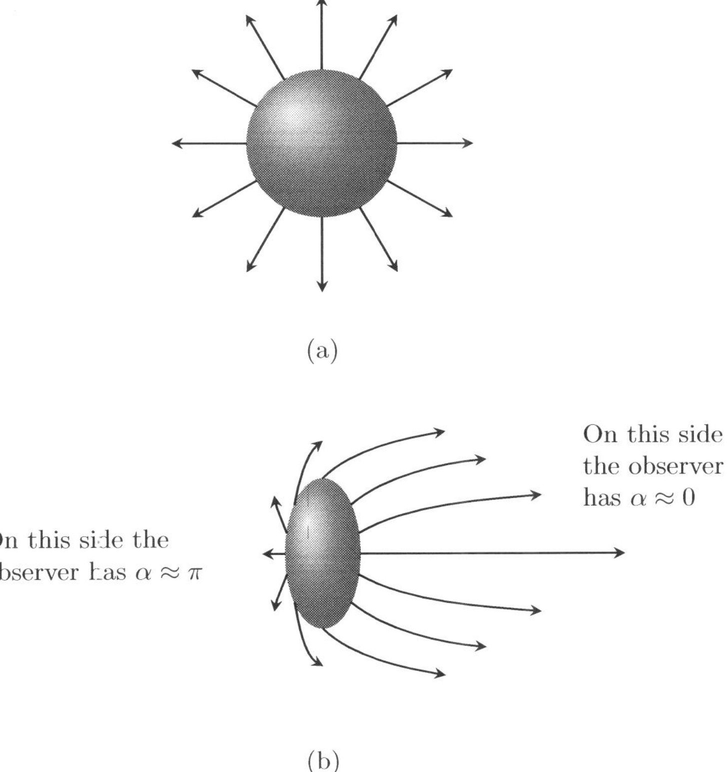

FIGURE 8.8. (a) A body, in its rest frame, uniformly radiating in all directions, with the momenta of 12 photons indicated by equal arrows. (b) A Lorentz contracted radiation picture, showing the same 12 photons as seen by the observer. In this case, $u \approx c$, and the intensity of light is highly concentrated in the forward direction.

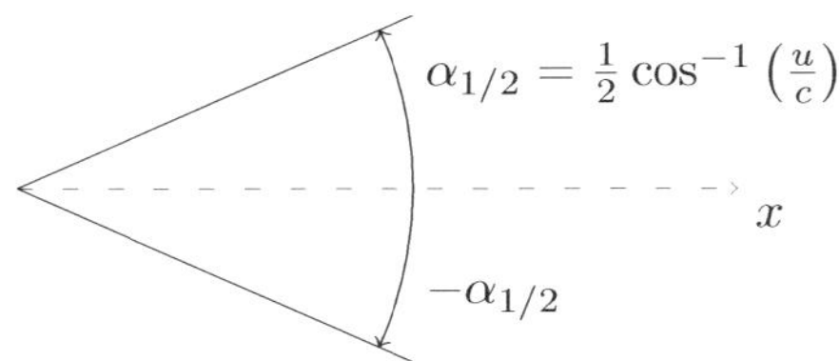

FIGURE 8.9. Cone for α as $u \to c$.

Another way to analyze this behavior is to realize that half of the photons emitted by the source in Fig. 8.6 are in a cone given by

$$|\alpha'| \leqq \frac{\pi}{2}, \tag{8.43}$$

or

$$0 \leqq \cos\alpha' \leqq 1, \tag{8.44}$$

which, from Eq. (8.19) becomes

$$0 \leqq \frac{\cos\alpha - \frac{u}{c}}{1 - \frac{u}{c}\cos\alpha} \leqq 1, \tag{8.45}$$

or since $1 - \frac{u}{c}\cos\alpha > 0 \quad \left(\frac{u}{c} < 1, \cos\alpha < 1\right)$

$$0 \leqq \cos\alpha - \frac{u}{c} \leqq 1 - \frac{u}{c}\cos\alpha. \tag{8.46}$$

The lower inequality in Eq. (8.46) implies that

$$\cos\alpha \geqq \frac{u}{c}, \tag{8.47}$$

and since $\cos\alpha \leqq 1$, we have

$$\frac{u}{c} \leqq \cos\alpha \leqq 1, \tag{8.48}$$

or

$$0 \leqq |\alpha| \leqq \cos^{-1}\left(\frac{u}{c}\right). \tag{8.49}$$

Eq. (8.49) means that, as $u \to c$, α lies in a very narrow cone centered about $\alpha = 0$; namely, the situation depicted in Fig. 8.9.

Lastly, using Eq. (8.18), we compare frequencies in the forward and backward directions (in the observational frame of reference), obtaining

$$\frac{\nu_\circ(\cos\alpha = 1)}{\nu_\circ(\cos\alpha = -1)} = \frac{\left(1 + \frac{u}{c}\right)}{\left(1 - \frac{u}{c}\right)}, \tag{8.50}$$

with the source frequency ν_s cancelling out from the numerator and denominator of the right-hand side of Eq. (8.50). Thus, we have established that

$$\nu_\circ(\text{forward direction}) \gg \nu_\circ(\text{backward direction}) \tag{8.51}$$

for $u \approx c$. This establishes part (b) of Fig. 8.8. We have not actually established any detailed analytic conclusions about the intensity of light; this is a separate, but related, problem. However, our analysis does imply that the intensity is much greater in the forward direction.

8.6. Problems

(1) For Fig. 8.3, for the direction of u shown, derive the result

$$\nu_\circ = \sqrt{\frac{(1 - u/c)}{(1 + u/c)}}\, \nu_s. \tag{8.13}$$

The derivation is the same as for Fig. (8.1), except we must remember that for the photon

$$P_x = -|\vec{p}| = -p = -\frac{h\nu}{c}.$$

(2) For Fig. 8.4, for the direction of u shown, derive the result

$$\nu_\circ = \sqrt{\frac{(1 + u/c)}{(1 - u/c)}}\, \nu_s. \tag{8.12}$$

(3) In Fig. 8.7, show that

$$\cos\alpha = \frac{\cos\alpha' + \frac{u}{c}}{\left(1 + \frac{u}{c}\cos\alpha'\right)} \tag{i}$$

$$\tan\alpha = \pm\frac{\sin\alpha'}{\gamma\left(1 + \frac{u}{c}\cos\alpha'\right)} \tag{ii}$$

$$\nu_\circ = \frac{\nu_s}{\left[\gamma\left(1 - \frac{u}{c}\cos\alpha\right)\right]}. \tag{iii}$$

Then, from Eq. (i), show that

$$\cos\alpha' + \frac{u}{c} = \cos\alpha + \frac{u}{c}\cos\alpha\cos\alpha'; \tag{iv}$$

or

$$\cos\alpha' = \frac{\cos\alpha - \frac{u}{c}}{1 - \frac{u}{c}\cos\alpha}. \tag{v}$$

(4) (a) For a distant star the frequency of the electromagnetic radiation emitted is 5×10^{16} hertz$(\sec^{-1})$, which is in the ultraviolet part of the spectrum. Assume the star is in a galaxy moving at $1/3$ of the speed of light (recession velocity) with respect to the Earth. What frequency will we observe here on earth?
(b) Is this a *red shift* or a *blue shift*?

(5) (a) The red shift of a very distant galaxy is given by $z = .75$. Find the velocity of recession of the galaxy. (You may find it convenient to express this velocity in units of the speed of light). Then, using Hubble's Law, find the distance of the galaxy in Mpc. (Hubble's constant is given by $H_\circ = 72$ (km/sec)/Mpc, where a Mpc $\rightarrow$ mega-parsec $= 3.26$ ly is a distance of measurement used by astronomers).
(b) From Hubble's Law, calculate in Mpc the distance beyond which galaxies move faster than the speed of light. Compare this value with the result of part (a).

Chapter **9**

Electrodynamics

Maxwell's equations in MKS (or SI) units are given by

$$\vec{\nabla} \cdot \vec{E} = \frac{\rho}{\epsilon_\circ} \tag{9.1a}$$

$$\vec{\nabla} \times \vec{E} = -\frac{\partial \vec{B}}{\partial t} \tag{9.1b}$$

$$\vec{\nabla} \cdot \vec{B} = 0 \tag{9.1c}$$

$$c^2 \vec{\nabla} \times \vec{B} = \frac{\vec{j}}{\epsilon_\circ} + \frac{\partial \vec{E}}{\partial t}, \tag{9.1d}$$

where $\vec{E}$ and $\vec{B}$ are, respectively, the electric and magnetic field vectors, $\epsilon_\circ$ is the permittivity of free space, ρ is the density of charge, and $\vec{j}$ is the vector current density. Also, we note the following: $c^2 = \frac{1}{\epsilon_\circ \mu_\circ}$, with $\mu_\circ$ the permeability of free space.

The equation for conservation of local charge is

$$\vec{\nabla} \cdot \vec{j} = -\frac{\partial \rho}{\partial t}, \tag{9.2}$$

which is also called the equation of continuity, while the force law is given by

$$\vec{F} = q(\vec{E} + \vec{v} \times \vec{B}), \tag{9.3}$$

for a charge q moving with a velocity $\vec{v}$ in electric and magnetic fields $\vec{E}$ and $\vec{B}$.

The electric and magnetic field vectors can be reexpressed in terms of the scalar potential ϕ and the vector potential $\vec{A}$:

$$\vec{B} = \vec{\nabla} \times \vec{A} \tag{9.4a}$$

$$\vec{E} = -\vec{\nabla}\phi - \frac{\partial \vec{A}}{\partial t}, \tag{9.4b}$$

and it is clear that Eqs. (9.4) automatically satisfy (9.1b) and (9.1c). The other Maxwell equations then become

$$-\nabla^2\phi - \frac{\partial}{\partial t}(\vec{\nabla} \cdot \vec{A}) = \frac{\rho}{\epsilon_\circ}, \tag{9.5a}$$

83

and

$$- c^2 \nabla^2 \vec{A} + c^2 \vec{\nabla}(\vec{\nabla} \cdot \vec{A}) + \frac{\partial}{\partial t} \vec{\nabla}\phi + \frac{\partial^2 \vec{A}}{\partial t^2} = \frac{\vec{j}}{\epsilon_\circ}. \tag{9.5b}$$

Eqs. (9.5) are complicated equations, which greatly simplify using the famous Lorentz gauge[*]

$$\vec{\nabla} \cdot \vec{A} + \frac{1}{c^2} \frac{\partial \phi}{\partial t} = 0, \tag{9.6}$$

giving for Eqs. (9.5)

$$\nabla^2 \phi - \frac{1}{c^2} \frac{\partial^2 \phi}{\partial t^2} = -\frac{\rho}{\epsilon_\circ} \tag{9.7a}$$

$$\nabla^2 \vec{A} - \frac{1}{c^2} \frac{\partial^2 \vec{A}}{\partial t^2} = -\frac{\vec{j}}{\epsilon_\circ c^2}. \tag{9.7b}$$

It is now convenient to define the following four-vectors

$$\mathcal{A} = \left(\vec{A}, \frac{1}{c}\phi \right) \tag{9.8}$$

$$\mathcal{J} = \left(\vec{j}, c\rho \right), \tag{9.9}$$

which allow Eqs. (9.7) to be reexpressed in four-vector form as

$$\Box^2 \mathcal{A} = -\frac{1}{\epsilon_\circ c^2} \mathcal{J}, \tag{9.10}$$

where

$$\Box^2 = \nabla^2 - \frac{1}{c^2} \frac{\partial^2}{\partial t^2} \tag{9.11}$$

is the d'Alembertian operator defined in Eq. (4.56). Also, the Equation of Continuity, Eq. (9.2) and the Lorentz gauge condition, Eq. (9.6), can be written, respectively, as the four-divergences

$$\partial \cdot \mathcal{J} = 0 \tag{9.12}$$

$$\partial \cdot \mathcal{A} = 0, \tag{9.13}$$

where, from Eq. (4.53), we have

$$\partial = \left(\vec{\nabla}, -\frac{\partial}{\partial(ct)} \right). \tag{9.14}$$

Notice, from Eqs. (9.12), (9.13), and (9.14), in the two four-divergences, there are two cancelling minus signs, one coming from ∂ and the other from the basic four-vector dot-product, Eq. (4.10).[†]

---

[*]General equations for gauge transformations are given and discussed in Appendix E. For our purposes in electrodynamics, it is important to emphasize that a gauge transformation involves changes in the relations of the derivatives of the vector and scalar potentials that leave $\vec{E}$ and $\vec{B}$ unchanged.

[†]It should be mentioned that for SR some books adopt the notation, in which a four-vector A can be labeled a contavariant vector A^μ, or a covariant vector A_μ, where μ can take on the values $1(x), 2(y), 3(z)$ or $0(t)$. In SR, the covariant vector is simply obtained from the contravariant vector by the replacement: $A_t = -A^t$. Then, the four-dimensional dot product in Eq. (4.10) can be written as: $A \cdot B = \sum_\mu A_\mu A^\mu$. This notation is sometimes useful in SR, especially in E&M theory, but because of the definition of the four-dimensional dot product, it is not necessary in SR. However, the contavariant/covariant notation does become necessary in the mathematics developed in GR, where one uses differential geometry, with a more complicated metric than Eq. (4.14). Also, this notation is often very useful in field-theoretical formulations involving SR, in which different kinds of particles and interactions are specified.

In problem #1, you are asked to verify that Eqs. (9.10), (9.12), and (9.13) involve four-dimensional dot products, so that these equations must all be invariant under Lorentz transformations. Note too that Eq. (9.10) can be written in component form

$$\Box^2 \mathcal{A}_\mu = -\frac{1}{(\epsilon_\circ c^2)} \mathcal{J}_\mu. \tag{9.15}$$

The left-hand-side of Eq. (9.15) is invariant under a Lorentz transformation, and the right-hand-side is a four-vector which must also be invariant under a Lorentz transformation. However, even though $\vec{A}$ and ϕ transform as a four-vector, the fields $\vec{E}$ and $\vec{B}$ defined in Eqs. (9.4) are not invariant under a Lorentz transformation. See Feynman [4, Page 26-9] for tables detailing the behavior of the $\vec{E}$ and $\vec{B}$ fields perpendicular and parallel to the direction $\vec{u}$ of the frame S' with respect to S.

We emphasize too that the elegant Eqs. (9.7) and (9.10) arise because we have imposed the Lorentz gauge Eq. (9.6) (or equivalently Eq. (9.13)). If we had imposed instead the Coulomb gauge

$$\vec{\nabla} \cdot \vec{A} = 0, \tag{9.16}$$

the beautiful equations (9.7) and (9.10) would not be valid. Note too that Eq.(9.16) is not Lorentz invariant. See Appendix E for more details about Gauge Invariance.

Also, we mention that the title of this chapter is *Electrodynamics*, not *Relativistic Electrodynamics*. The basic equations discussed are the usual Maxwell equations (9.1), along with Eqs. (9.2)–(9.14), all of which are part of the normal formulation of classical electromagnetic theory. Einstein recognized that, especially after using the Lorentz gauge, Eq. (9.6), the resulting Eqs. (9.7) are exceptionally beautiful and elegant. Thus, one of Einstein's goals in developing SR was to save classical electrodynamics, even if it meant overthrowing Newton's laws. In 1905 this was a shock to many physicists since Maxwell's equations were only about 40 years old, while Newton's laws went back more than 200 years. In this context, note the title of Einstein's original 1905 paper published in "Annalen der Physik": "On the Electrodynamics of Moving Bodies", which emphasizes electromagnetism, *not* mechanics.

9.1. Problem

(1) Verify Eqs. (9.10), (9.12) and (9.13). In particular, show that these equations involve four-dimensional dot-products. See subsection 4.8.

Chapter **10**

Superluminal Motion; Real, or an Optical Illusion?

In Chapter 8, subsection 8.4, we discussed the superluminal expansion of space, an effect from General Relativity, whereby space is not forbidden from expanding faster than the speed of light. This phenomenon is real and does not violate SR. In this chapter we shall discuss several other cases of superluminal motion, one of which is an optical illusion.

Technically, superluminal motion refers to an object that apparently moves faster than the speed of light. The questions could be raised: what object and in *what medium?* In subsection 10.1, refraction occurs due to the difference between the speed of light in a substance like water compared to the speed of light in air (which we can consider practically to be the same as a vacuum). Also, in subsection 10.1 we discuss Cherenkov radiation which occurs when the speed of an object in a particular medium is less than the speed of light in a vacuum but greater than the speed of light in that medium. This mechanism for Cherenkov radiation is exactly the same as that in the sonic boom, the shock wave produced by an airplane breaking the sound barrier.

In subsection 10.2, we discuss the differences between the phase and group velocities of a particle wave (a "wavicle") moving in a vacuum, and we show from SR that the real physical velocity, the group velocity, is always less than c (the velocity of light in a vacuum), but the phase velocity of the wave exceeds c. (Note; for a photon, the group and phase velocities are the same and are equal to the speed of light.) Then, in subsection 10.3 we consider the apparent motion of certain, very distant astronomical objects, moving in space.

10.1. Index of Refraction and Cherenkov (or its variation Cerenkov) Radiation

The first, trivial example of what could be described as superluminal motion occurs for *refraction* at the boundary of a *dispersive medium*, and air. (See Fig. 10.1.) No matter which direction you approach the surface boundary, the ray is bent, but the speed of light in the dispersive medium is always less than the speed of light in the air (which could be considered the same as a vacuum). (For simplicity we consider a flat boundary, but similar behavior occurs if the boundary is curved.) The dispersive medium has an index of refraction, n, that depends on frequency, with

$$n > 1 \qquad (10.1)$$

for a range of frequencies, typically in the optical region. Then, the speed of light in such a medium is given by

$$\text{speed of light}_{\text{medium}} = v_l = \frac{c}{n} < c. \qquad (10.2)$$

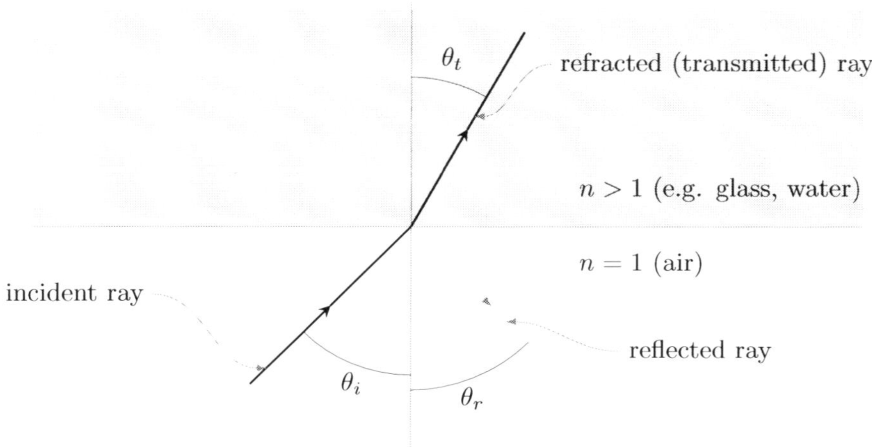

FIGURE 10.1. Reflection and refraction at a plane surface boundary between air and a dispersive medium.

Assuming that the light originates in air, at the boundary of the (flat) surface, the law of reflection dictates that

$$\Theta_i = \Theta_r, \tag{10.3}$$

while the law of refraction, *Snell's Law*, states that

$$n \sin \Theta_t = \sin \Theta_i. \tag{10.4}$$

Eqs. (10.3) and (10.4) are well-known results in geometric optics, and these equations can also be derived using the wave theory of light presented in Chapter 9. Moreover, Eqs. (10.3) and (10.4) can be obtained from a variational method, known as *Fermat's Principle of Least Time* (1650). Now, refraction, a simple case of subluminal motion, gives us a mechanism for studying the following important example of superluminal motion.

An interesting application of Eq. (10.2) occurs in elementary-particle physics, in a phenomenon known as *Cherenkov radiation*, for which we have

$$c > v_p > v_l, \tag{10.5}$$

where v_p, the speed of a particle in the medium, is greater than $v_l = c/n$, the speed of light in the medium (but less than the speed of light in a vacuum). This behavior can be seen in Fig. 10.2, in which we see that

$$\cos \Theta = \frac{v_l t}{v_p t} = \frac{v_l}{v_p} = \frac{c}{n v_p}, \tag{10.6}$$

or

$$\cos \Theta = \frac{1}{\beta n} \qquad \left(\beta = \frac{v_p}{c} < 1 \right) \tag{10.7}$$

$$> \frac{1}{n}.$$

However, from Eq. (10.1) we have $\frac{1}{n} < 1$, so that

$$\frac{1}{n} < \cos \Theta < 1, \tag{10.8}$$

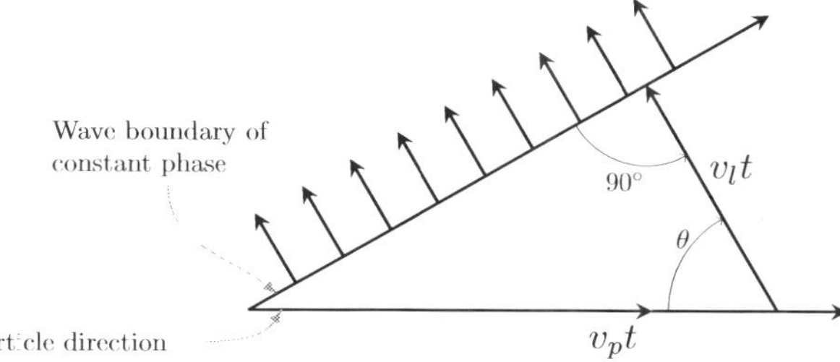

FIGURE 10.2. Cherenkov cone of radiation, where t is the time elapsed.

which defines a cone about the direction of the particle.

Thus, the particle radiates light (called Cherenkov radiation) in a cone about the direction of the particle, as shown in Fig. 10.2. This behavior is exactly the same as the *shock wave* experienced in a sonic boom, when a plane travels faster than the speed of sound.

In nuclear physics, Cherenkov radiation is observed in water-modulated nuclear reactors, with a signature of beautiful blue light surrounding the core of the reactor. Also, Cherenkov radiation is predicted in the deep underground "swimming pool" detectors in which physicists have searched for proton decay. However, proton decay has not been observed, although the detectors have successfully been used for various new developments in neutrino physics.

10.2. Group and Phase Velocities

For wave motion, we now calculate both the phase velocity, which is the rate at which the phase of the wave propagates through space, and the group velocity, which is the overall rate at which the envelope of the wave moves through space. We also use elementary wave (quantum) mechanics to formalize the mathematics of the wave motion. Then, for waves identified as particles, the group velocity is interpreted as the particle velocity.

First, construct some elementary quantum mechanical operations, such as the quantum mechanical momentum operator in the x-direction

$$\mathcal{P}(\text{operation in the } x\text{-direction}) = \mathcal{P}_x(\text{operator}) = -i\hbar\frac{\partial}{\partial x}, \tag{10.9}$$

where

$$\hbar = \frac{h}{2\pi} \tag{10.10}$$

and h is Planck's constant. We want to find the eigenvalues and eigenfunctions of this operation. Then, we examine

$$\mathcal{P}_x(\text{operator})e^{ikx} = \hbar k e^{ikx}, \tag{10.11}$$

so that $P = \hbar k$ is the eigenvalue corresponding to the eigenfunction $\phi = e^{ikx}$, and Eq. (10.11) can be rewritten as

$$\mathcal{P}_x(\text{operator})\phi = \mathcal{P}(\text{number})\phi = p\phi , \tag{10.12}$$

where $p(\text{number}) = p = \hbar k$. Also, from Quantum Mechanics, recall the de Broglie relation

$$p = \frac{h}{\lambda} = \frac{2\pi\hbar}{\lambda} = \hbar k, \tag{10.13}$$

or

$$k = \frac{2\pi}{\lambda}. \tag{10.14}$$

Next, construct the quantum mechanical energy operator

$$\mathcal{E}(\text{operator}) = i\hbar\frac{\partial}{\partial t}, \tag{10.15}$$

or

$$\mathcal{E}(\text{operator})e^{-i\omega t} = Ee^{-i\omega t} = \hbar\omega e^{-i\omega t}, \tag{10.16}$$

so that the eigenvalue is $\hbar\omega$ corresponding to the eigenfunction $e^{-i\omega t}$. Also, recall Eq. (7.33) for a photon:

$$E = h\nu = 2\pi\hbar\nu. \tag{10.17}$$

Comparing Eq. (10.16) and (10.17) we find that

$$\omega = 2\pi\nu. \tag{10.18}$$

Thus, for free-particle motion in one dimension (x) we use a general plane-wave eigenfunction, constructed from Eqs. (10.11) and (10.16)

$$\Psi(x,t) = Ae^{i(kx-\omega t)}, \tag{10.19}$$

where A is a real normalization constant. Next, we want to construct the simplest possible type of *wave packet* by considering the sum of two functions, each like Eq. (10.19), and with k and ω values (or equivalently p and E values) very close to one another, so we take the *superposition*

$$\bar{\Psi}(x,t) = \Psi_1(x,t) + \Psi_2(x,t) = A\left[e^{i(k_1x-\omega_1t)} + e^{i(k_2x-\omega_2t)}\right], \tag{10.20}$$

with

$$\omega_1 \approx \omega_2 \qquad (\text{but } \omega_1 \neq \omega_2) \tag{10.21a}$$
$$k_1 \approx k_2 \qquad (\text{but } k_1 \neq k_2) . \tag{10.21b}$$

In physics, we often take the real parts of complex functions, so we define

$$\mathcal{X}(x,t) = Re[\bar{\Psi}(x,t)] = A[\cos(\omega_1 t - k_1 x) + \cos(\omega_2 t - k_2 x)], \tag{10.22}$$

which gives a very simple wave packet. (See Fig 10.3.) Then, from a well-known trigonometric identity, Eq. (10.22) becomes

$$\mathcal{X}(x_1,t) = 2A\cos\left[\left(\frac{\omega_1 + \omega_2}{2}\right)t - \left(\frac{k_1 + k_2}{2}\right)x\right] \times \cos\left[\left(\frac{\omega_1 - \omega_2}{2}\right)t - \left(\frac{k_1 - k_2}{2}\right)x\right] \tag{10.23}$$

Now, adopt the notation

$$\omega \equiv \frac{\omega_1 + \omega_2}{2} \approx \omega_1 \approx \omega_2 \tag{10.24a}$$
$$k \equiv \frac{k_1 + k_2}{2} \approx k_1 \approx k_2 \tag{10.24b}$$
$$\Delta\omega \equiv \omega_1 - \omega_2 \tag{10.24c}$$
$$\Delta k \equiv k_1 - k_2, \tag{10.24d}$$

giving

$$\mathcal{X}(x,t) = 2A\cos\left(\omega t - kx\right)\cos\left(\frac{\Delta\omega}{2}t - \frac{\Delta k}{2}x\right). \tag{10.25}$$

The first factor in in Eq. (10.25) is a wave with velocity

$$v = \nu\lambda = \frac{2\pi\nu}{(2\pi/\lambda)} = \frac{\omega}{k}, \tag{10.26}$$

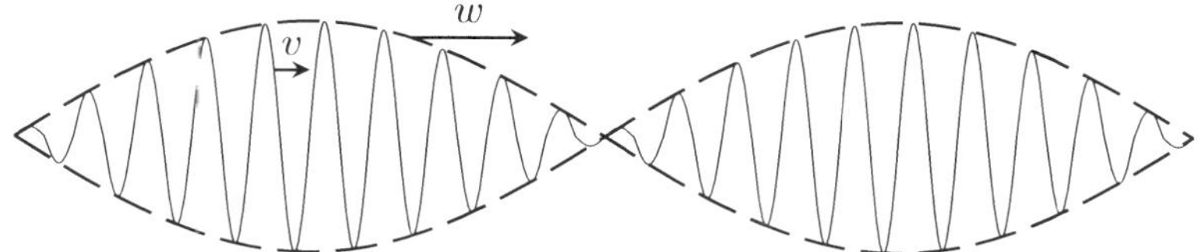

The wave motion consists of a rapid oscillation, enclosed by a more slowly varying envelope:

The solid line above progresses with the *phase velocity*, $v = \frac{E}{p}$, which is (usually) greater than the group velocity, $w = \frac{\mathrm{d}E}{\mathrm{d}p}$, of the envelope. Thus, the wave train appears to be constantly moving through the slower envelope, without distorting its shape.

FIGURE 10.3. Phase velocity v and group velocity w of wave packets

which is the *phase velocity*

$$v = \frac{\hbar\omega}{\hbar k} = \frac{E}{p}. \tag{10.27}$$

The second factor is a much more slowly varying function, representing the envelope of the wave packet in Fig. 10.3, giving a velocity

$$w = \frac{\Delta\omega/2}{\Delta k/2} \to \frac{\mathrm{d}\omega}{\mathrm{d}k}, \tag{10.28}$$

taking the limit of differentials, which can be rewritten as

$$w = \frac{\mathrm{d}(\hbar\omega)}{\mathrm{d}(\hbar k)} = \frac{\mathrm{d}E}{\mathrm{d}p}, \tag{10.29}$$

a velocity which is called the *group velocity*, corresponding to the true, or physical, velocity of the particle.

As we see in Problem #2 for this chapter, for a relativistic particle

$$v > c, \tag{10.30}$$

while

$$w < c, \tag{10.31}$$

with

$$vw = c^2. \tag{10.32}$$

Thus, indeed the phase velocity can exceed c, but the real physical velocity is w, the group velocity. The phase velocity is not a real velocity, so there is no violation of SR.

10.3. (Apparent) Superluminal Expansion of "Clouds" (or "Blobs") From Distant Radio Sources

In Ref. 8, page 61 and in Ref. 18, pages 348–351, there are discussions of the movement of "clouds" or "blobs", tens of thousands light years across, which emerge from radio sources; for example, the radio source 3C345. These blobs give the appearance of having a speed, in a transverse plane perpendicular to the line of sight, which exceeds the speed of light. This behavior arises as follows: if you take the apparent perpendicular distance in the transverse plane and divide by the

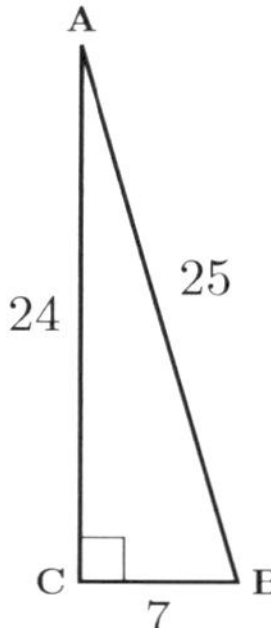

FIGURE 10.4. The 24–7–25 right triangle, ABC

elapsed time between observations, you can naïvely obtain transverse velocities (speeds) that are greater than c. Are these velocities real? The answer, as we shall demonstrate, is that this is a fallacious calculation, and that the real effect is due to the fact that the blobs are moving almost straight at us, with speeds very close to, but less than, c.

Before proceeding further, we give a definition from astronomy and some calculations. A light year is a measure of distance; namely, the distance that light will travel in one year (in a vacuum), or

$$1 \text{ light year (ly)} = c \times \text{year} \tag{10.33}$$

$$= 2.998 \times 10^8 \frac{\text{m}}{\text{sec}} \times 1 \text{ year}$$

$$= 9.454 \times 10^{17} \text{cm}$$

$$= .3060 \text{ Pc (Parsec)},$$

or

$$\text{distance (in ly)} = c \times (\text{number of years}) \tag{10.34}$$

and

$$\text{velocity (in units of } c) = \frac{\text{distance in light years}}{\text{number of years required to observe the distance}}. \tag{10.35}$$

It will also be helpful, as an example, to use the 24-7-25 right triangle, shown in Fig. 10.4.

First, see Fig. 10.5, in which it is assumed that the "blob" moves with the speed of light along the hypotenuse of the right triangle. From Eq. (10.34) and (10.35), we find that the perpendicular velocity is

$$\frac{7 \text{ ly}}{(t_\circ + 1 - t_\circ) \text{ years}} = \frac{7}{1} \frac{\text{ly}}{\text{y}} = 7c$$

where $t_\circ$ is the time at which the blob left the source. More realistically, in Fig. 10.6, we assume that the blob moves with a speed $v = \frac{24}{25}c = .96c$ along the hypotenuse. Now the perpendicular speed is given by

$$\sim \frac{7\text{ly}}{(t_\circ + 2 - t_\circ)\text{y}} = \frac{7\text{ly}}{2\text{y}} = 3.5c.$$

Thus, we see that you can observe an apparent transverse velocity $> c$ by having the blob move approximately toward us, with a speed less than c, but close to c. For a long while, these transverse speeds fooled some astronomers, who believed that transverse speeds greater than c were possible.

92

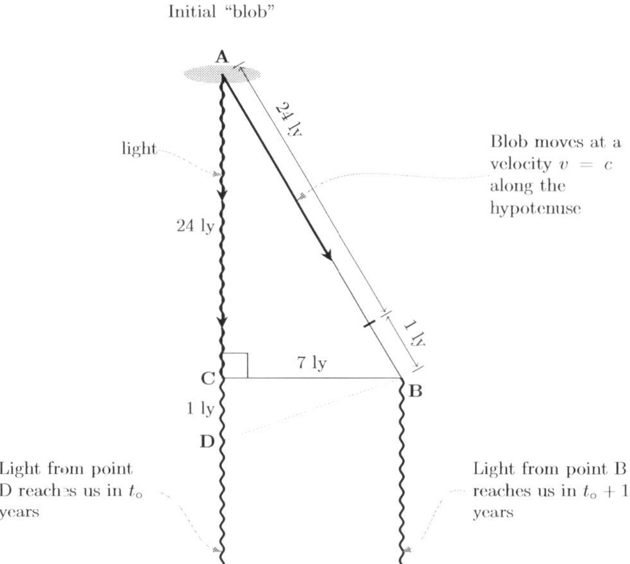

FIGURE 10.5. Assuming that the blob moves at the speed of light along the hypotenuse of the 24–7–25 (ABC) right triangle (not drawn to scale). In year zero, light is emitted from point A and reaches us $25 + t_\circ$ years later. 25 years after emitting light at point A, the blob reaches point B, but the light emitted at point B reaches us only one year after we observed the light emitted at point A.

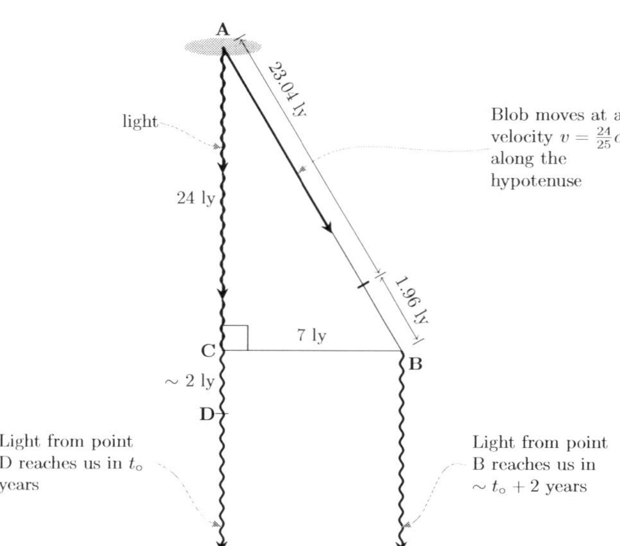

FIGURE 10.6. Assuming that the blob moves at $\frac{24}{25}$ the speed of light along the hypotenuse of the 24–7–25 (ABC) right triangle (not drawn to scale). In year zero, light is emitted from point A and reaches us $25 + t_\circ$ years later. 25 years after emitting light at point A, the blob reaches point B, but the light emitted at point B reaches us ~two years after we observed the light emitted at point A.

$$\textbf{10.4. Problems}$$

(1) (a) Prove that the *vector* operator for momentum is
$$\vec{P}(op) = -i\hbar\vec{\nabla},$$
with eigenvalue $\hbar\vec{k}$ and eigenfunction $e^{i\vec{k}\cdot\vec{r}}$.

   (b) Show that the four momentum operator can be expressed as
$$P(op) = -i\hbar\partial$$
where ∂ is defined in Eq. (4.53).

(2) (a) For a relativistic particle, show that
$$v = \frac{E}{p} = \sqrt{\frac{m_\circ^2 c^4 + p^2 c^2}{p}} = c\sqrt{1 + \frac{m_\circ^2 c^2}{p^2}} > c$$

and
$$vw = c^2,$$

so that
$$w = \frac{c^2}{v} = c\left(\frac{c}{v}\right) < c$$

   (b) For a photon show that
$$v = w = c.$$

Chapter 11

A Brief, Qualitative Introduction to General Relativity (GR)

11.1. Description and Features of GR

First, we make some remarks about the differences between Einstein's SR and GR theories. The word "special" in SR refers to the requirement that, at least in the main formulation of the theory, one deals with reference frames moving with a constant relative velocity with respect to one another. On the other hand, in GR the word "general" indicates that one deals with accelerated relative motion. In particular, in GR we will be especially concerned with the generalization of the acceleration arising from the Newtonian gravitational force*.

SR gives better accuracy than Newtonian mechanics, especially when dealing with objects moving at high speeds (say, moving at a fair fraction of the speed of light). GR improves the accuracy of Newton's Law of Gravity, especially when dealing with very massive objects. However, in the limits of relatively small speeds and small masses, Newton's Law of Motion and Newton's Law of Gravity are still quite good approximations to reality. For example, Newtonian calculations in the solar system are quite reliable. (Two exceptions: precession of the perihelions of the inner planets, especially Mercury, and GPS calculations, which extensively use both SR and GR.)

GR determines the curvature of SpaceTime, which—in turn—governs how objects move. See Fig. 11.1. Thus, there is a *self-consistency* of concepts here: mass and energy tell space how to curve and the curvature—in turn—governs the motion of matter (and energy). These two effects

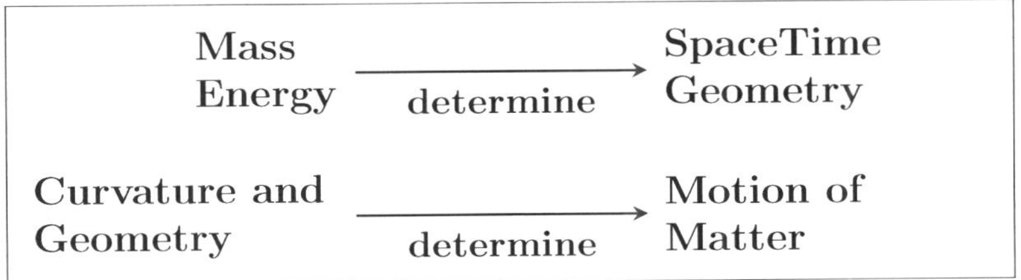

FIGURE 11.1. Self-consistency in GR

---

*However, we recall that, in subsections 3.4 and 4.5, we mentioned that in SR the acceleration of a particle in a particular Lorentz frame of reference can be simulated by the introduction of a series of infinitesimal inertial frames, each with its own infinitesimal light cone, along the particle's world line in SpaceTime.

are embodied very nicely in Einstein's equations for GR:

$$R^{\mu\nu} - \frac{1}{2}g^{\mu\nu}R = -\frac{8\pi G}{c^4}T^{\mu\nu}, \tag{11.1}$$

where μ, ν are SpaceTime variables, $R^{\mu\nu}$ is a geometric factor, involving derivations of the metric term g, R is the *scalar curvature* (formally called the trace of the so-called Ricci tensor), $T^{\mu\nu}$ is an energy-momentum tensor and G is the universal gravitational constant. The left-hand-side of Eq. (11.1) expresses the dynamic nature of the geometry; the right-hand-side energy expresses the energy/momentum dynamics. A more qualitative way of seeing these effects is often presented in texts as the "bowling ball/rubber sheet" analogy. In this model, a bowling ball distorts a rubber sheet, which is a representation of the influence of mass (the bowling ball) on SpaceTime (the rubber sheet). Then, the indentation of the rubber sheet allows a marble to roll on its distorted surface, showing how objects move under the influence of curvature. Although easy to grasp, this analogy has been criticized by some authors for being an oversimplification.

We see then that Fig. 11.1 shows there is an intimate connection between SpaceTime Geometry and Mass/Energy. Note that, strictly speaking, gravity is not a "force", like the electromagnetic, strong and weak forces since the gravitational particle trajectories, known as geodesics, arise from a classical variational principle and—in fact—the same variational method developed in Chapter 6.

Particles follow *Geodesic Curves* in SpaceTime, responding to the curvature, not to the influence of a "force".

FIGURE 11.2. Geodesics as SpaceTime Trajectories

Isaac Newton was very concerned about a particular problem concerning his theory of gravity. For his force, he was bothered by the unphysical *action at a distance* between two masses:

$$F = \frac{mMG}{r^2}, \tag{11.2}$$

where F is the gravitional force, m and M are the masses attracted to one another, separated by a distance r. If we move either mass, then the force at the other changes instantaneously. This feature of the classical theory contradicts SR, which dictates that nothing can move faster than the speed of light, which was one of the important factors in motivating Einstein to develop in 1916 GR, a new theory of gravity. In GR the *effective force* arises from the gravitational curvature of the SpaceTime geometry, so there no need to invoke an instantaneous infinite velocity for the transmission of the force.

11.2. The Equivalence Principal and Tidal Forces

Consider Eq. (11.2) as Newton's Second Law:

$$F = m_i a = \frac{m_g M_g G}{r^2}, \tag{11.3}$$

where m_i is the inertial mass and m_g and M_g are gravitational masses. It was realized very early that

$$m_i = m_g, \tag{11.4}$$

so that the two masses cancel out from Eq. (11.3). This explains why all objects experience the same acceleration in a gravitational field, regardless of their masses. Recall the famous experiment attributed (perhaps as an apocryphal story) to Galileo that different mass objects, dropped from the

Leaning Tower of Pisa, fall to Earth with the same acceleration. (Ideally, one should perform the acceleration experiment in a vacuum since obviously in air there an extra force, the wind resistance, that will alter the behavior of a particular body according to its mass.)

Eq. (11.4) is a truly amazing result! It is not at all obvious. In fact, recall that the acceleration of an electrically charged object does depend on both its charge *and* its mass.

Einstein elevated this to a fundamental principal of GR, which is known as the *Equivalence Principal*, stating that:

The effect of gravity is equivalent to, or indistinguishable from, the effect of an acceleration (and vice versa).

FIGURE 11.3. The Equivalence Principal

For example, suppose that you are in a space ship orbiting Earth at a constant velocity in a spherical orbit, as in Fig. 11.4, so that you are in "free fall". From Eqs. (11.3) and (11.4), there is an acceleration given by (using also the acceleration equation for centrifugal motion)

$$a = \frac{v^2}{r} = \frac{MG}{r^2}, \tag{11.5}$$

with

$$v = \sqrt{\frac{MG}{r}}. \tag{11.6}$$

Observers in the spaceship are said to be in free fall, which is one of several examples of such motion[†] The point is that the force of gravity has the same effect on the observers as motion in a uniformly rotating system of coordinates moving with the spaceship. If you did not know that you

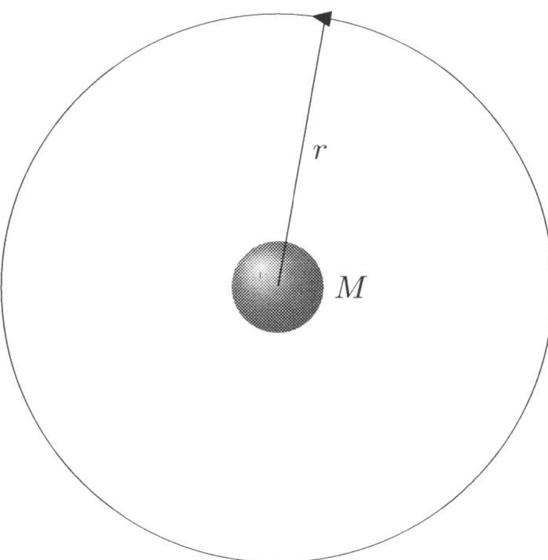

FIGURE 11.4. Free Fall, with M the mass of the Earth and r the distance from the space ship to the center of the Earth.

[†]Another example of free fall is an elevator whose cables have been cut and which is plummeting toward the Earth. All of us would rather experience free fall in the spaceship than in the elevator.

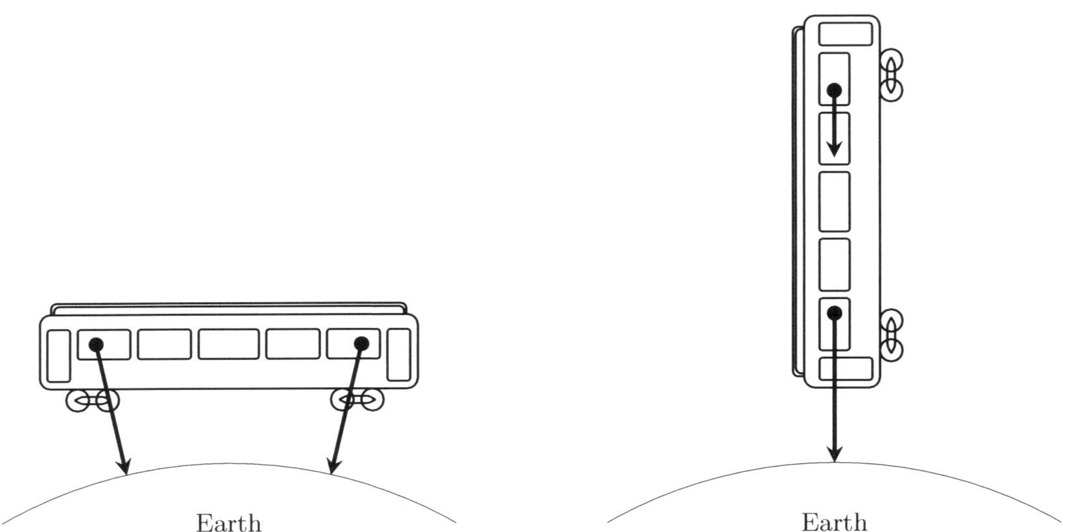

FIGURE 11.5. Two orientations of Einstein's old fashioned railway coach in free fall. The arrows show the magnitude and direction of the gravitational force on that part of the railway coach.

are in a spaceship rotating about the Earth, you could not distinguish your motion from that of a uniform circular motion.

However, in applications of the Equivalence Principal, it is important to keep in mind that it is an exact statement only in the limit of an infinitesimally small object. Equivalently, the principal is exact for an extended object only if *tidal forces* are eliminated. Tidal forces are illustrated in Fig. 11.5, which demonstrates the concept using Einstein's famous thought-experiment railway coach. You see that for an extended object these gravitational forces differ in magnitude and direction, depending on what part of the object you are considering.

It can be shown that all tidal forces are second-order effects, and gravity is indistinguishable from an acceleration in first order. Thus, while there are no exact gravitational inertial frames, there exist approximate ones to which we can apply a version of the Equivalence Principal.

11.3. Confirmed Predictions of GR

11.3.1. Gravitational Time Dilation, or, Equivalently, the Gravitational Red Shift.
Previously, in Chapter 3 we discussed time dilation (leading to the Twin Paradox) in SR. Here we discuss a related type of time dilation that occurs in GR due to gravitational effects.

It has been argued these two types of time dilation are identical, an argument which is based on the Equivalence Principal relating gravitational effect to acceleration. (Recall that in Chapter 3 that the Twin Paradox resolution involved taking into account the motion of the twin who had experienced accelerations and decelerations. However, in GR time dilation occurs when only classical gravitational acceleration is present, without any other kind of mechanical acceleration - as in SR.) We want to keep the two effects seperate since they appear to arise from different mechanisms; one, in SR, from motion, the other, in GR, from a strong gravitational field.

Also, in SR time dilation can only be appreciable if an object is travelling at a speed that is a fair fraction of the speed of light. Similarly, the time dilation in GR is only measurable if the mass causing the effect is large (say, at least as large as the Earth, or better, the Sun).

11.3.1.1. *Statement of the effect.* Clocks are observed to run *more slowly* in a *strong gravitational field* (as measured by an observer in a *weaker* gravitational field)*. For example, an atomic clock kept at the National Institute of Standards and Technology (NIST) in Boulder, Colorado (1650 meters above sea level) gains about 5 microseconds a year relative to an identical clock kept at the Royal Greenwich Observatory (25 meters above sea level). As will be discussed, differences in clocks can be expressed as differences in frequencies. Thus, in 1960 Pound and Rebka measured the change in frequency of a 14.4 KeV γ-ray (emitted by the radioactive decay of ^{59}Fe) between two different levels, separated by 22.6 m. Such incredible precision is possible because of the Mössbauer effect, which gives spectral lines of very precisely defined frequencies in radiation emitted from an atomic nucleus in a crystal.

The time dilation effect has also been detected in the light emitted near white-dwarf stars, giving a gravitational red shift.[†] The most dramatic case of gravitational time dilation, according to theory, would be the light emitted near a black hole (BH), for which the slowing down of time is really phenomenal. For example, any object with mass moving toward the event horizon of a BH would be seen moving ever more slowly as it encounters more and more the intense gravitational field of the BH. Yet, according to theory, the object itself experiences a normal time difference (in its frame of reference) as it approaches and passes the event horizon. (See Chapter 13.)

11.3.1.2. *Calculation of the effect.* We now develop some mathematics for the theory of gravitational time dilation, or the gravitational red-shift. First, because the time difference Δt is proportional to a period of oscillation, we have

$$\Delta t \propto P = \nu^{-1} \propto \lambda, \tag{11.7}$$

where P is the period of the electromagnetic radiation, ν is its frequency, and λ is its wave length. (Recall $c = \lambda\nu, P = \nu^{-1}$.) Thus, if Δt is more, ν is less and λ is greater, compared to the electromagnetic radiation emitted in the absence of a gravitational field.

Physically, the effect arises because a photon has to climb out of a strong gravitational field (with mass M), losing energy as it moves to a lower gravitational field, thereby (from $E = h\nu$) reducing its frequency, as in Fig. 11.6.

You can understand this effect, in a heuristic way, using Quantum Mechanics, SR, and Newtonian gravity. We write the energy of the photon as:

$$E_\nu = \{\text{"intrinsic energy" from Quantum Mechanics}\} + \{\text{gravitational potential energy}\}, \tag{11.8}$$

or

$$E_\nu(r) = h\nu - \left(\frac{h\nu}{c^2}\right)\frac{MG}{r}, \tag{11.9}$$

where, from Eqs. (5.32) and (7.33), the effective photon mass is given by

$$m_{\text{photon}}^{(\text{eff})} = \frac{E_{\text{photon}}}{c^2} = \frac{h\nu}{c^2}, \tag{11.10}$$

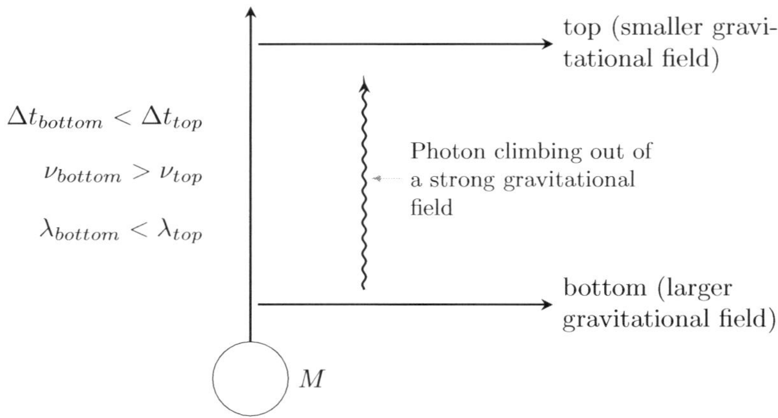

FIGURE 11.6. Illustrating the Gravitational Red Shift

where h is Planck's constant and G is the universal gravitational constant. We assume that energy is conserved in going from one radius to another, so that

$$h\nu_1 \left[1 - \frac{G}{c^2} \frac{M}{r_1} \right] = h\nu_2 \left[1 - \frac{G}{c^2} \frac{M}{r_2} \right], \tag{11.11}$$

and

$$\frac{\nu_2}{\nu_1} = \frac{1 - \frac{G}{c^2} \frac{M}{r_1}}{1 - \frac{G}{c^2} \frac{M}{r_2}}, \tag{11.12}$$

where the photon, in climbing out of the (classical) potential well, gains gravitational potential energy, but loses intrinsic energy. Thus, we see that if

$$r_1 < r_2 \qquad (\text{i.e., } r_1 \text{ is closer to } M \text{ than } r_2),$$

then

$$1 - \frac{G}{c^2} \frac{M}{r_1} < 1 - \frac{G}{c^2} \frac{M}{r_2},$$

and

$$\frac{\nu_2}{\nu_1} < 1 \qquad (\text{or } \nu_2 < \nu_1),$$

the gravitational red-shift. Thus, we have

$$\text{If } r_1 < r_2, \text{ then } \nu_1 > \nu_2, \tag{11.13}$$

the red-shift formula. On the other hand,

$$\text{If } r_1 > r_2, \text{ then } \nu_2 > \nu_1, \tag{11.14}$$

the blue-shift formula, which would occur for a photon going from a weaker gravitational field to a stronger one. Eqs. (11.13) and (11.14) can be derived rigorously using the formalism of GR.

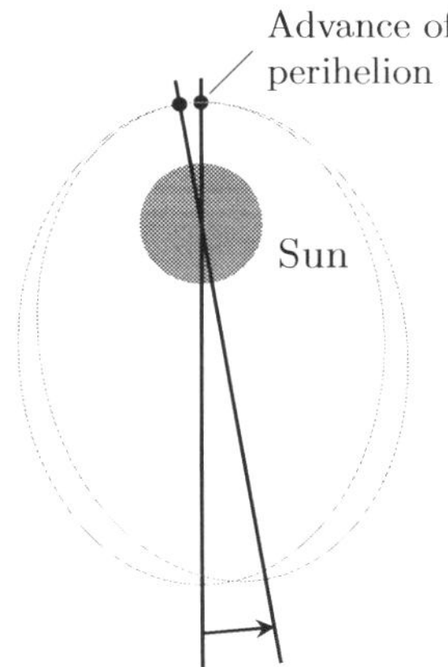

FIGURE 11.7. Precession of the Perihelion of Mercury (exaggerated view).

11.3.2. Precession of the Perihelion of Mercury. The orbits of the planets in our solar system are not fixed ellipses. Instead, after a 360° rotation, an orbit of a planet does not continue on its previous path, so it deviates ever so slightly from an ellipse, as we see in Fig. 11.7 for the planet Mercury. The perihelion is the distance of closest approach to the Sun, and it is observed to "precess", or rotate, about a line through its orignal path. The effect has been measured for all of the inner planets, but it is most pronounced for the planet Mercury.

The total precession of the perihelion of Mercury was long known to be 574" of arc per century, where " = 1 sec = 1/3600 of a degree. Most of this effect can be accounted for by the Newtonian gravitational forces of the other planets and the asteroids in the solar system. However, a stubborn $\sim 43"$ of arc per century can not be accounted for by Newtonian physics, but can be correctly calculated by GR, as can be seen in a simple exercise described in Ref. [52], Project C.

It was said that this result excited Einstein very much.

11.3.3. Bending of star light passing near the Sun (or other massive body). Fig. 11.8 shows the effect. When the Earth is on one side of Sun, the star is seen in its *usual* (or *true*) position. However, on the other side of the Sun, the light from the star is bent by the Sun, inferring an *apparent* position of the star. Moreover, during a total eclipse of the Sun (with the Moon lying between the Earth and the Sun), this apparent position can be directly observed since light from the star is not overwhelmed by the Sun's luminosity.

More precisely, a (partial or total) eclipse of the Sun occurs when the Moon passes between the Earth and the Sun, with all three bodies in conjuction (aligned). In a total eclipse, the disk of the Sun is fully obscured by the Moon. In a partial eclipse, only part of the Sun is obscured. Thus, only in a total eclipse can stars behind the Sun become visible for a short period of time[‡].

Such an effect was first observed in a total eclipse of the Sun in 1919 by Sir Authur Eddington. It was said that in 1919 Eddington had totally accepted GR and was intent on proving the theory by his observations, which entailed him travelling to the island of Principe, off of the coast of Equatorial Guinea in West Africa, where the total eclipse could be observed.

[‡]Also, the effect of a total eclipse can only be observed in a well-defined geographical section of the Earth.

The bending occurs because energy is mass (see Eq. (11.10).) Thus, light experiences the gravitational influence of the Sun, galaxies, etc. It should be mentioned too that there is a classical bending, which is calculated to be about half as much as that predicted by GR.

The bending of electromagnetic radiation has also been observed for other cases; e.g., in gravitational lensing in which light, emitted by distant quasars, is bent by intervening galaxies.

11.3.4. The slowing of light near the Sun. In addition to predicting the bending of electromagnetic radiation by a massive body, GR also predicts that near a massive body the speed of light slows down. Thus, I. I. Shapiro and collaborators performed a series of experiments in which they bounced radio waves between Earth and Mercury, Earth and Venus, and Earth and Mars, when these planets were situated on the opposite side of the Sun as shown in Fig. 11.9.

Because the velocity of light is slowed when it passes the Sun, Shapiro and his collaborators observed time delays in their measurements of the reflected radio waves. (The delay is with respect to the time it would take in normal flat SpaceTime; i.e., the transit time, without the Sun present.)

11.3.5. Black Hole (BH). A BH is predicted by GR. However, since any object (a particle with mass or a photon) can not escape the clutches of the BH if it is closer to the BH than (or at) its event horizon

$$r_{EH} = 2\frac{GM}{c^2}, \tag{11.15}$$

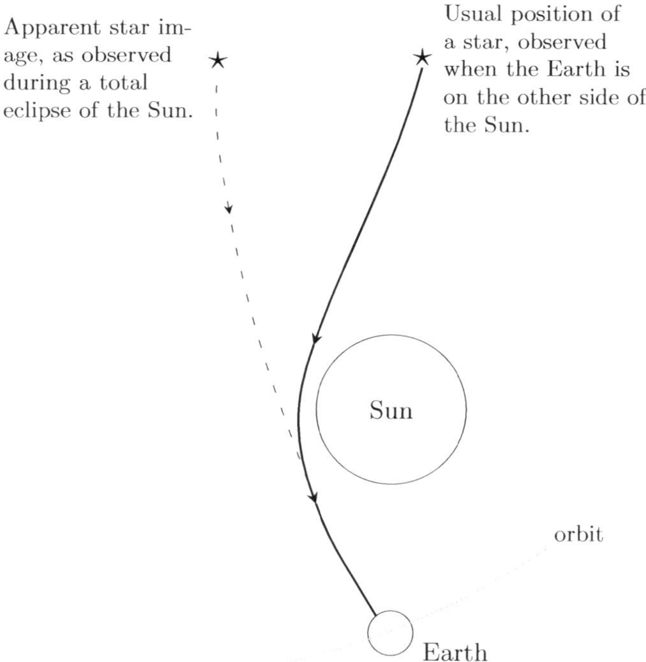

FIGURE 11.8. Bending of star light passing near the Sun, as observed in a total eclipse of the Sun.

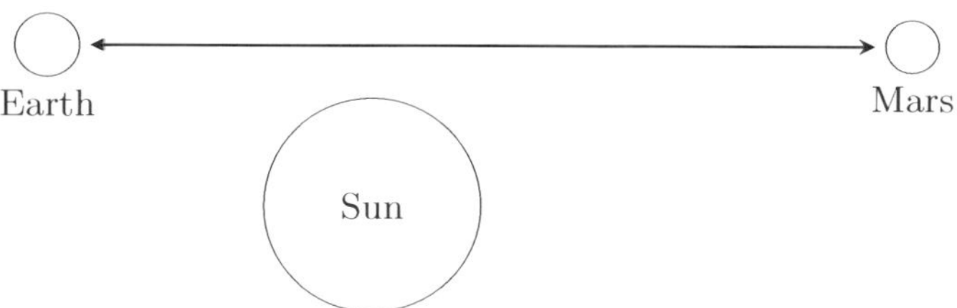

FIGURE 11.9. Round-trip radio signal between the Earth and Mars. Similar transits were observed for the planets Mercury and Venus.

where M is the mass of the BH*. This means that we can not directly observe light emanating from within or on the horizon of a BH. Thus, BH's have very strange properties which depend critically on the observer making the measurements. (In this sense, both SR and GR are very much like Quantum Mechanics, which is well-known to be very observer-dependent. This point is discussed in more detail in Chapter 13.) Also, we note that the event horizon, given in Eq. (11.15), can be simply obtained from the expression for the escape velocity: $v^2 = \frac{2GM}{r}$, derived from Newton's laws. Then simply set the escape velocity to c, the velocity of light, and solve for r (the event horizon).

However, there are means by which a BH can be observed. For example, in a binary star system, in which only one member of the binary is a BH, the motion of the visible member of the binary can infer properties of its partner. Also, in a binary star system, you can detect a BH by examining the X-rays from the region outside of the event horizon. Material from the visible companion star will literally flow or swirl into the BH, giving off energetic X-rays. (This flow of gas near a BH is called the *accretion disk*.)

It is belived that very massive BHs exist at the centers of many galaxies (including our own Milky Way galaxy). Also, the current theory of quasars (quasi-stellar radio sources) is that such objects pertain to galaxy formation in the very early Universe, with the galaxy being "powered" by a massive BH.

11.3.6. The Big Bang theory of the evolution of the Universe. Many books have been written about the good empirical evidence for the Big Bang cosmology[†] (See particularly Reference 50). The theoretical underpinning of the Big Bang is GR.

11.3.7. The Global Positioning System. Finally, it should also be mentioned that an enormous practical application of GR is the famous Global Positioning System (GPS). In the GPS, "triangulation" with respect to at least 4 of 24 satellites in circular (12 hour) orbits around the Earth can be used to determine a terrestrial location (longitude, latitude, and elevation). To obtain adequate accuracy, GPS receivers must take into account time dilation effects from both SR and GR: from SR, that time passes more slowly on the rapidly moving satellites than at the less rapidly moving GPS receiver; and from GR that time passes more slowly at the receiver (closer to the Earth's mass) than at the satellites in orbit.

---

*There is no radiation from a BH, except for the very special (quantum mechanical) Hawking radiation, whose existence has yet to be confirmed in our present Universe.

†For many years in the twentieth century the Steady State Theory was a competing cosmology, but by the end of the century, the Big Bang had decisively won the battle.

(1) (a) From Eq. (11.11), for the case $\nu_1 \approx \nu_2 \approx \nu$

$$\frac{\nu_2 - \nu_1}{\nu_1} \approx \frac{GM}{c^2}\left(\frac{1}{r_2} - \frac{1}{r_1}\right). \tag{i}$$

(b) Then, in the small mass limit

$$\frac{M}{\text{radius}} \ll \frac{c^2}{G}, \tag{ii}$$

prove that (i) is equivalent to

$$\frac{\nu_2 - \nu_1}{\nu_1} \approx \frac{\left[1 - \frac{2MG}{c^2}\frac{1}{r_1}\right]^{1/2}}{\left[1 - \frac{2MG}{c^2}\frac{1}{r_2}\right]^{1/2}} - 1, \tag{iii}$$

or

$$\frac{\nu_2}{\nu_1} = \sqrt{\frac{\left(1 - \frac{2MG}{c^2}\frac{1}{r_1}\right)}{\left(1 - \frac{2MG}{c^2}\frac{1}{r_2}\right)}}, \tag{iv}$$

a completely general formula in the full GR theory. For the purposes of discussion, assume that the two radii in Eq. (iv) are arbitrary radii outside the horizon of the BH. Next, note that as $r_2 \to \frac{2G}{c^2}M = r_{EH}$ (the event horizon from Eq. (11.15)), $\frac{\nu_2}{\nu_1} \to \infty$, and as $r_1 \to r_{EH}$, $\frac{\nu_2}{\nu_1} \to 0$, both of which imply that light cannot escape from a BH. Why?

God used beautiful mathematics in creating the world.

Paul Dirac

I have an equation; do you have one too?

Paul Dirac on meeting Richard Feynman

Chapter 12

The Dirac Equation

Perhaps the most important application of SR is the merging of the theory with Quantum Mechanics, giving the famous Dirac equation (1927). The Dirac equation was exceptionally important for field theory and other developments in theoretical physics during the remainder of the twentieth century. The Dirac equation is fundamental for understanding the behavior of electrons, so it is also of immense practical value. Thus, we now give a short chapter demonstrating how the Dirac equation is derived, concluding with two subsections discussing the Lorentz covariance of the equation.

12.1. The Main Equation in Energy-Momentum Space

In what follows, we deal mostly with a particle in free space, with no interactions.

In the early twentieth century, it was realized that, due to Quantum Machanics and special relativity, the laws of physics had to satisfy the famous Klein-Gordon equation:

$$P \cdot P c^2 \Psi = (\vec{p} \cdot \vec{p} c^2 - E^2)\Psi = -m_\circ^2 c^4 \Psi, \tag{12.1}$$

where P is the relativistic four momentum, and E and $\vec{p}$ are the relativistic energy and vector momentum, respectively, which in operator form act on the quantum mechanical wave function Ψ (see Eqs. (5.36)–(5.38)); $m_\circ$ is the mass of the particle under consideration (usually the electron).

However, from very general principles of Quantum Mechanics, including the superposition principle, Paul Dirac in 1927 realized that it should be possible to express the fundamental relativistic equation for an electron (and by inference any spin -1/2 particle), in terms of a linear superposition of the energy and momentum operators. Moreover, putting energy and momentum on the same footing gives one the opportunity to develop a covariant equation (i.e., with the operators obeying a symmetry under a Lorentz transformation), a feature that we will discuss at the end of this chapter. Dirac also recognized that the non-relativistic Schrödinger equation was not suitable as a relativistic equation since it is linear in energy but quadratic in momentum, while the Klein-Gordon equation, (12.1), an appropriate relativistic equation, has serious problems with probability. (It was clear though that the "square" of any linear equation had to satisfy the Klein-Gordon equation). Specifically, Dirac postulated an equation of the form

$$(E - c\vec{\alpha} \cdot \vec{p})\Psi = +\beta m_\circ c^2 \Psi, \tag{12.2}$$

107

for which it was discovered that the four expansion functions $\vec{\alpha}$ and β are each 4×4 matrices, whose dimensions have important physical implications (to be discussed in due course). Then, the wave function Ψ is a 4-component column vector.

We now perform an operation to effectively "square" the Dirac equation in order to compare with the Klein-Gordon equation, a procedure that will give us conditions which the α and β matrices must satisfy. First rewrite Eq. (12.2) as

$$\left[-c\vec{\alpha} \cdot \vec{p} + E - \beta m_\circ c^2 \right] \Psi = 0, \tag{12.3}$$

which we multiply on the left by the operator

$$(c\vec{\alpha} \cdot \vec{p} + E + \beta m_\circ c^2), \tag{12.4}$$

giving

$$\begin{aligned}
\{ &-c^2 (\vec{\alpha} \cdot \vec{p})(\vec{\alpha} \cdot \vec{p}) + E^2 - \beta^2 m_\circ^2 c^4 \\
&+ Ec(\vec{\alpha} \cdot \vec{p} - \vec{\alpha} \cdot \vec{p}) \\
&+ Em_\circ c^2 (\beta - \beta) \\
&- m_\circ c^3 (\vec{\alpha} \cdot \vec{p}\beta + \beta\,\vec{\alpha} \cdot \vec{p}) \} \Psi = 0,
\end{aligned}$$

or

$$\begin{aligned}
\{ &c^2 (\vec{\alpha} \cdot \vec{p})(\vec{\alpha} \cdot \vec{p}) - E^2 + \beta^2 m_\circ^2 c^4 \\
&+ m_\circ c^3 \left[(\vec{\alpha} \cdot \vec{p})\beta + \beta(\vec{\alpha} \cdot \vec{p}) \right] \} \Psi = 0.
\end{aligned} \tag{12.5}$$

We write the $\vec{\alpha}$ and β matrices in the form given in Eq. (12.5) since in general they do not commute with one another, i.e.,

$$\begin{aligned}
\alpha_i \alpha_j &\neq \alpha_j \alpha_i & (i, j = 1, 2, 3;\ i \neq j) \\
\alpha_i \beta &\neq \beta \alpha_i. & (i = 1, 2, 3)
\end{aligned}$$

We next rewrite the first and fourth terms of Eq. (12.5) as:

$$c^2 (\vec{\alpha} \cdot \vec{p})(\vec{\alpha} \cdot \vec{p}) = c^2 \sum_{i,j=1}^{3} \alpha_i \alpha_j p_i p_j \tag{12.6a}$$

$$m_\circ c^3 \left[(\vec{\alpha} \cdot \vec{p})\beta + \beta(\vec{\alpha} \cdot \vec{p}) \right] = m_\circ c^3 \sum_{i=1}^{3} (\alpha_i \beta + \beta \alpha_i) p_i. \tag{12.6b}$$

Then, in order to satisfy Eq. (12.1), Eqs. (12.5) and (12.6) demand that

$$\begin{aligned}
\{\alpha_i, \alpha_j\} &= \alpha_i \alpha_j + \alpha_j \alpha_i = 0; & (i, j = 1, 2, 3;\ i \neq j) && \text{(12.7a)} \\
\{\alpha_i, \beta\} &= \alpha_i \beta + \beta \alpha_i = 0; & (i = 1, 2, 3) && \text{(12.7b)} \\
\alpha_i^2 &= 1 & (i = 1, 2, 3) && \text{(12.7c)} \\
\beta^2 &= 1, & && \text{(12.7d)}
\end{aligned}$$

where the brackets $\{\}$ denotes anti-commutation.

In order to satisfy Eqs. (12.7), non-unique forms of the 4×4 matrices $\vec{\alpha}$ and β are given by

$$\alpha_i = \left(\begin{array}{ccc} 0 & \vdots & \sigma_i \\ \cdots\cdots\cdots \\ \sigma_i & \vdots & 0 \end{array} \right) \qquad (i = 1, 2, 3) \qquad (12.8\text{a})$$

$$\beta = \left(\begin{array}{ccc} I & \vdots & 0 \\ \cdots\cdots\cdots \\ 0 & \vdots & -I \end{array} \right), \qquad (12.8\text{b})$$

where the dotted lines denote separation of the 4×4 matrices into block 2×2 matrices, I is the unit 2×2 matrix, and the σ_i's are the famous 2×2 Pauli matrices, defined as:

$$\sigma_1 = \begin{pmatrix} 0 & 1 \\ 1 & 0 \end{pmatrix} \qquad \sigma_2 = \begin{pmatrix} 0 & -i \\ i & 0 \end{pmatrix} \qquad \sigma_3 = \begin{pmatrix} 1 & 0 \\ 0 & -1 \end{pmatrix}. \qquad (12.9)$$

Note too that

$$\sigma_i^2 = I \qquad\qquad (i = 1, 2, 3) \qquad\qquad (12.10\text{a})$$

$$\{\sigma_i, \sigma_j\} = 0. \qquad\qquad (i, j = 1, 2, 3; \ i \neq j) \qquad\qquad (12.10\text{b})$$

From Eqs. (12.8) and (12.10), Eqs. (12.7) follow. Also, we see that Eqs. (12.7a), (12.7c), and (12.10) can be written in a more compact form.

$$\{\alpha_i, \alpha_j\} = 2\delta_{ij} \qquad\qquad (i, j = 1, 2, 3) \qquad\qquad (12.11\text{a})$$

$$\{\sigma_i, \sigma_j\} = 2\delta_{ij}. \qquad\qquad (i, j = 1, 2, 3) \qquad\qquad (12.11\text{b})$$

The arguments establishing that $\vec{\alpha}$ and β are 4×4 matrices and that Ψ is a 4-component column vector are given in many texts, such as Refs. 70, 74 and 75. Then, the dimensionality of 4 has the following interpretation. Two of the components of the space refer to the intrinsic spin of the electron and the other two refer to "negative-energy" states, which by using modern field theory can be reinterpreted to establish the existence of the positron, which is the anti-particle of the electron.

Thus, with his famous equation, Dirac was credited with establishing i) the existence of the intrinsic spin $(= \frac{1}{2}\hbar)$ of the electron, and ii) of predicting the existence of the positron, which was discovered by Carl Anderson in 1932 in his cloud chamber experiments involving cosmic rays. Dirac won the Nobel prize in Physics in 1933 and Anderson, the prize in 1936.

The limitations of a single-particle theory (like the Dirac equation) in relativistic Quantum Mechanics has been eloquently expressed by Steven Weinberg in his essay "How Great Equations Survive" (Ref. 76). In a slightly nuanced tribute to Dirac's famous equation, which was intended as a relativistic extension of the ordinary Schrödinger equation, Weinberg notes, in the context of modern field theory, that

> "The trouble with all of this is that there is no relativistic quantum theory of the sort for which Dirac was looking. The combination of relativity and Quantum Mechanics inevitably leads to theories with unlimited numbers of particles." [i.e.: field theories]

However, Weinberg goes on to note that there are cases where a good approximation can be obtained from the action of a field operator on states containing a single electron (e.g., the case of the fine structure of hydrogen). Weinberg then states that, for such a case,

"Because the equation for the electron field operator is mathematically the same as Dirac's equation for his wave function, the results of this calculation turn out to be the same as Dirac's."

The coincidence of a Dirac-like equation mimicing a field operator equations is not confined to fermions. It appears that almost all of the elementary particle bosons satisfy such equations (which are basically Maxwell-generalized equations), with the possible exception of the spin-zero Higgs boson. Thus, in refs 71 and 72, it is shown, respectively, how to develop Dirac-like equations for massless bosons (the photon and graviton) and the spin-one, massive bosons (the $W^{\pm}$ and Z° particles in ElectroWeak theory).

12.2. The Dirac Equation, with Derivatives

The quantum mechanical operators for the energy-momentum four-vector are given by

$$P = \left(\vec{p}, \frac{E}{c}\right) = -i\hbar\left(\vec{\nabla}, -\frac{1}{c}\frac{\partial}{\partial t}\right) = -i\hbar\partial, \qquad (12.12)$$

where $\hbar = \frac{h}{2\pi}$ (h is Planck's constant) and the ∂ four-vector is defined in Eq. (4.53). The Dirac equation, (12.2), can then be written in SpaceTime coordinates as

$$\left(i\hbar\frac{\partial}{\partial t} + ci\hbar\,\vec{\alpha}\cdot\vec{\nabla}\right)\Psi = \beta m_{\circ}c^2\Psi. \qquad (12.13)$$

12.3. The Covariant Form of the Dirac Equation

Define the following 4×4 matrices

$$\gamma_t = \gamma_{\circ} = \beta \qquad\qquad (12.14a)$$
$$\gamma_k = \beta\alpha_k, \qquad\qquad (k = 1, 2, 3) \qquad\qquad (12.14b)$$

from which you can easily show that

$$\{\gamma_u, \gamma_v\} = \gamma_u\gamma_v + \gamma_v\gamma_u = -2g_{uv}, \qquad (u, v = t, 1, 2, 3) \qquad (12.15)$$

where the g_{uv} are matrix elements of the metric tensor defined in Eq. (4.14). Another way of writing $\vec{\gamma}$ is:

$$\vec{\gamma} = \begin{pmatrix} 0 & \vdots & \vec{\sigma} \\ \cdots & \cdots & \cdots \\ -\vec{\sigma} & \vdots & 0 \end{pmatrix}. \qquad (12.16)$$

Substituting Eqs. (12.14) into Eqs. (12.3) and (12.13), we obtain the following:

$$\left[E - c\gamma_t(\vec{\gamma}\cdot\vec{p}) - \gamma_t m_{\circ}c^2\right]\Psi = 0 \qquad (12.17a)$$

and

$$\left[i\hbar\frac{\partial}{\partial t} + ci\hbar\gamma_t\vec{\gamma}\cdot\vec{\nabla} - \gamma_t m_{\circ}c^2\right]\Psi = 0, \qquad (12.17b)$$

which can, respectively, be reexpressed as

$$\left[E\gamma_t - c(\vec{\gamma}\cdot\vec{p}) - m_{\circ}c^2\right]\Psi = 0 \qquad (12.18a)$$

$$\left[\gamma_t\left(i\hbar\frac{\partial}{\partial t}\right) + ci\hbar(\vec{\gamma}\cdot\vec{\nabla}) - m_{\circ}c^2\right]\Psi = 0. \qquad (12.18b)$$

Feynman introduced a "dagger" notation to simplify these equations. His dagger is defined as

$$\mathcal{C}\!\!\!/ = -\gamma_t C_t + \vec{\gamma} \cdot \vec{C}, \tag{12.19a}$$

where C is a general four-vector, giving for instance

$$\nabla\!\!\!/ = \partial\!\!\!/ = \gamma_t \frac{\partial}{\partial(ct)} + \vec{\gamma} \cdot \vec{\nabla}. \tag{12.19b}$$

Then, Eqs. (12.18) become

$$(P\!\!\!/ + m_\circ c)\Psi = 0 \tag{12.20a}$$

$$(i\hbar\nabla\!\!\!/ - m_\circ c)\Psi = 0, \tag{12.20b}$$

so that Eqs. (12.20) are consistent with Eq. (12.12).

In order to treat an interacting particle (say, with the electromagnetic field), we must use the gauge prescription in Eq. (E.6), so that Eqs. (12.20) become:

$$(P\!\!\!/ - q\mathcal{A}\!\!\!/ + m_\circ c)\Psi = 0 \tag{12.21a}$$

$$(i\hbar\partial\!\!\!/ + q\mathcal{A}\!\!\!/ - m_\circ c)\Psi = 0, \tag{12.21b}$$

where q is the charge of the particle and $\mathcal{A}$ is given by Eq. (9.8) (for an interaction with the electromagnetic field).

Note that the Feynman dagger for a four-vector in Eqs. (12.19) gives a relativistically covariant quantity, as we see from Eqs. (4.12). Thus, Eqs. (12.20) and (12.21) are the covarient expressions for the Dirac equation.

12.4. Closing Remarks

From the point of view of rigor, the above presentation is a bit of a swindle. We have been following the treatment in Ref. 75, pages 17–20, a subsection which is entitled "Covariant Form of the Dirac Equation". However, in the next subsection of Ref. 75 on pages 21-26, the authors formulate the full relativistic invariance of the Dirac equation, in which they demonstrate the Lorentz transformation properties of the gamma matrices, defined in Eqs. (12.14). This presentation, known as "Pauli's fundamental theorem", dictates that one must perform the same similarity transformation on each of the 4×4 gamma matrices, after which one obtains a second set of matrices that are manifestly four vectors*. Then, the relativistic invariance of the Dirac equation is completely demonstrated, and it is accurate to claim that Eqs. (12.20) and (12.21) are the true covariant forms of the Dirac equation. Recall that the α and β matrices, whose properties were originally chosen to give agreement between the Dirac equation and the Klein-Gordon equation, are used to form the γ matrices. Then, the γ matrices are fundamental in establishing the covariance of the Dirac equation. Since covariance was not built into the α and β matrices, this behavior seems almost magical.

We conclude with an additional quote from Weinberg's essay:

> "...When an equation is as successful as Dirac's, it is never simply a mistake. It may not be valid for the reason supposed by its author, it may break down in new contexts, and it may not even mean what its author thought it meant. We must continually be open to reinterpretations of these equations. But the great equations of modern physics are a permanent part of scientific knowledge, which may outlast the great cathedrals of earlier ages."

---

*This derivation is quite complicated and beyond the scope of this book.

Chapter **13**

The Relativity of Measurements in Physics

Since this is a book on relativity, we will conclude with a short summary of the relativity of measurements in modern theories of physics—particularly measurements made in Quantum Mechanics, SR, and GR. We begin with (i) Quantum Mechanics followed by (ii) SR, and (iii) GR.

(i) In quantum theory, if you make a precise measurement of a position of a particle, you will know nothing about its velocity, or momentum. This result arises from the famous uncertainty principle:

$$\Delta x \, \Delta p_x \geq \sim \hbar \tag{13.1}$$

where Δx means the uncertainty in x-position and Δp_x, the uncertainty in x-momentum; $\hbar$ is Planck's constant h divided by (2π). Similar relations apply to other *conjugate variables*, such as y and p_y as well as angles and angular momentum. In fact, the uncertainty principle is a type of relation covered in the more general complementarity principle, another example of which is the wave-particle duality, which dictates that quantum objects have both wave-like and particle-like properties. Thus, if you structure an experiment to measure its particle-like properties, you may lose all information about its wave-like properties, which may have to be determined by another experiment. For example, in the photoelectric-effect experiment, it was established that the photon is a particle. On the other hand, the double-slit experiment clearly establishes the wave-like nature of light. Also, the parable of *Schrödinger's cat* reminds us again of the role of the observer. The cat has a 50% chance of being alive and a 50% chance of being dead until an observer opens the box to determine the cat's true fate.

(ii) In SR, we have studied the role of the moving frame (S'), moving with a velocity u with respect to a stationary frame (S). Of course, there is no such thing as an absolute stationary frame since one of the main results of SR is that there is no preferred frame of reference (e.g., the nineteenth century ether frame). Thus, if we think of the Earth as our stationary frame, we have to realize that the Earth is moving with respect to the Sun, but the Sun is moving with respect to the center of the Milky Way galaxy, the Milky Way is moving with respect to our local group of galaxies, which—in turn—is moving with respect to other groups of galaxies, etc. (Apparently, all of the galaxy clusters are moving toward a mysterious object called the "Great Attractor", at least part of which may lie beyond the visible Universe.) But for now, for the purpose of discussion, we take our stationary frame S to be that of the Earth. We know that, relative to our stationary frame, a ruler at rest in the moving frame S' will appear to us to be shortened, and events that are simultaneous in our frame will not be simultaneous in S' (and vice versa). Moreover, a clock at rest in S' appears to us to be running more slowly. Also, if S' is moving away from us and if S' contains a source of light with a particular frequency ν', a receiver in our frame will measure a lower frequency than ν'. (See the relativistic Doppler shift, discussed in Chapter 8.) Then, in studying elementary-particle reactions, like the threshold cases described in section 7.2, it is convenient to define inertial systems that have convenient properties. For example, in the center-of-mass (CM) system, the total relativistic momentum of the colliding particles is zero, while in the laboratory (Lab) system of coordinates, in the prescattering reaction, one of the initial particles is at rest. Since the CM and Lab frames are related by a Lorentz transformation, there is an invariant valid in either frame, so by evaluating this invariant in both frames, we can uniquely solve the scattering problem. Thus, once again we see that different frames of reference (or, if you like, different observers) see different physics, but the properties of the physics in these frames can be related giving us useful information.

Perhaps the most spectacular example of the relativity of measurements in SR is time dilation, which is embodied in Eqs. (3.20), (3.22), and (3.23). The proper time, τ, in these equations refers to a clock at rest in an object that has a travel velocity v with respect to a stationary frame, which we can take to be the Earth. The time recorded in the stationary frame is denoted by t. In the usual twin paradox, the moving object—usually postulated to be a spaceship—travels at a velocity that is a fair fraction of the speed of light. Then, the time elapsed between two SpaceTime events on the clock affixed to the moving object records a much smaller value than that recorded by a clock on Earth. The closer the velocity of travel is to the velocity of light, the greater the elapsed times between the two clocks. We normally focus on spaceships in this problem, but it is also instructive to take the limit of a particle that travels at the speed of light and has zero mass (e.g., a photon). (See section 5.5.) In such a case, no matter what the time increment recorded on Earth, Δt, the proper time recorded on the photon, $\Delta\tau$, is *identically zero*, a result that follows from the time-dilation equations. This means that a photon will always travel instantaneously between *any two events* in SpaceTime! Thus, while light observed here on the stationary Earth from a distant galaxy may take billions of years to reach us, the *photons in that light experience instantaneous travel*. It is important to consider a world consisting of an ensemble of massless particles, all moving at the speed of light, a scenario that existed just after the Big Bang, but before the Universe cooled and before spontaneous symmetry breaking gave rise to Higgs-like particles. In that stage of the universe, every particle experienced instantaneous travel. Then, it is well known that the Higgs particle, pertaining to the ElectroWeak interaction, was responsible for giving mass to the known leptons and quarks, as well as to W^+, W^-, and, Z°, three of the ElectroWeak carrier particles. (Interestingly, however, the fourth ElectroWeak carrier particle, the photon, remained massless in the Higgs process.)

(iii) One of the best books on GR is Ref. [52] *Exploring Black Holes, Introduction to General Relativity*, by E.F. Taylor and J. A. Wheeler (T&W), which is basically pitched for undergraduates who know classical mechanics, SR, and calculus (both differential and integral). This excellent book contains very lucid explanations, and there are many humorous passages. The book adopts a simple metric approach to GR, but does not contain any sophisticated mathematics, with almost all exercises and problems amenable to solution using basic algebra and calculus. T&W give the reader great insights into all aspects of GR, from static to rotating Black Holes (BHs), from objects with mass to light, to all of the many examples of confirmed predictions of GR mentioned in Chapter 11, and eventually conclude with a chapter on elementary cosmology. One of the exceptional features of the T&W book is that it discusses the relativity of observations of a BH, which for many purposes in the book can be any massive object. There are at least three possible observers of a massive object: (1) a *shell observer* on a spherical shell outside of the object; (2) a *free-float observer* on, say, an unpowered space ship plunging in free fall toward the object; and (3) the so-called *Schwarzschild bookkeeper*, who patches together all of the measurements of observers (1) and (2). It should be noted too that the *Schwarzschild observer* is not a single person, but is really the collection of all of the data pertaining to SpaceTime, giving the *Schwarzschild metric* (which was studied in Problem #2 for Chapter 6). This data, however, allows us to assess what an *observer at infinity* would see. In project B of T&W, the authors also develop a special metric for a free-float observer who starts out at rest an infinite distance away, travels to the BH, and passes through its event horizon, eventually plunging toward certain oblivion at the center of the BH*. For a free-float observer (not a shell observer), humans can exist very close to the horizon of a BH.

In general, time increments recorded by a far-away observer are considerably greater than the corresponding increments recorded on a shell close to the massive object, which behavior is the basis for gravitational time dilation (or, equivalently, the gravitational red shift) and which was discussed in subsection 11.3.1. On the other hand, a far-away radial spatial increment is proportionally less than the corresponding shell-observer radial increment (reminding us of the Lorentz contraction in SR). However, perhaps the most striking behavior of the relativity of observation occurs for events near, at, and below the event horizon of a BH. Observing a free-fall spaceship, starting from rest at infinity and approaching the horizon, a nearby shell observer will see the speed of the spaceship increase, finally reaching the speed of light at the horizon. In contrast, a distant observer will see the speed of the spaceship decrease and reach zero at the horizon; however, the time that it takes the distant observer to see the spaceship reach the horizon is an infinite interval. On the other hand, for a BH having a mass such that tidal forces are relatively minimal at the event horizon (true of many known or suspected BHs), the plunging space ship passes through the horizon, with nothing untoward happening. In particular, the astronauts inside experience no sudden shudder, jolt, or tear upon crossing the horizon. Then, after passing through the event horizon of such a BH, the astronauts can often survive for a long time before suffering their final fate. (See the footnote below.) For example, if the free-fall spaceship crossed the event horizon for a static BH with a mass of 10^{10} sun masses, the astronauts inside could live another 43 hours! T&W suggest that, during the final interval, the astronauts aboard could take notes and record data on their laptops, with the motto "Publish *and* Perish"!

There is also a relativity of observation for light signals in GR. T&W obtain general expressions for the radial and transverse speeds of light for different observers. The magnitude of the speed of light is equal to c for all shell and free-float observers, but for a far-away observer the magnitude of the speed of light decreases near a massive object. Thus, unusual behavior with electromagnetic radiation, particularly with respect to distant observers, has been observed: the bending of starlight around the Sun (subsection 11.3.3), the slowing of radar near the Sun (subsection 11.3.4), and Einstein rings (images of a distant radio galaxy being bent by an intervening gravitating object; T&W, Projects D.7–D.10). In Chapter 5,

---

*T&W, in Project B, Queries 15–21, there is an estimate of the time during which a free-float observer approaching the center, or singularity, of a static BH will experience discomfort due to the tidal forces, a time they facetiously call "the ouch time", which they calculate to be 0.314 seconds for any static BH (independent of its mass). Since medical doctors estimate that pain signals travel through the nerves at the speed of approximately one meter per second, it is clear that death due to the singularity of a BH is exceeding fast and, thus, very merciful compared to most other ways of dying.

T&W give different views of light images of a BH and distant stars for a free-falling observer and a shell observer, as seen at different shells outside of a BH, with all data being obtained from the Schwarzschild bookkeeper.

Strictly speaking, the above results for BH's apply to nonrotating holes or to objects rotating very slowly (e.g., the Earth and the Sun). The results for rotating BH's are very similar, with one important exception: according to T&W a distant observer sees a spaceship, starting from rest at infinity and plunging toward the BH, and then ending up spiralling around the horizon approaching the speed of light. (Note: the radial component of the velocity approaches zero at the "static-limit radius", which is above the horizon.) Then, once again those on the spaceship experience nothing unusual in passing through the horizon (if the BH has minimal tidal effects at the horizon).

In summary, it is clear that measurements in the modern specialties of physics can depend very strongly on the nature of the observation and/or the motion/spatial location of the observer. Thus, in general, physical measurements are indeed relative to the characteristics of the observation or the nature of the observer. Indeed, this is relativity!

Important Milestones in the History of Electricity and Magnetism

This survey of electricity and magnetism (E&M) includes important discoveries regarding light and other components of the electromagnetic spectrum, as well as the interaction of the electromagnetic field with charged particles, (such as QED, which is the relativistic quantum mechanical theory of the interaction of the electromagnetic force with charged particles). Also, the survey includes some history of the weak interactions since the final culmination of that work was the ElectroWeak theory, which involves a unification of the weak interactions and QED. Moreover, one of the purposes of this appendix is to demonstrate that fundamental work in E&M is intimately connected to many well known discoveries in other fields of physics, and indeed in chemistry, biology, and geology. The emphasis is almost exclusively on the fundamental rather than on the practical or commercial. For example, we will discuss extensively discoveries involving the wave and particle (photon) nature of light, but we will devote almost no space to the practical development of telephones, radio, electronics, x-ray machines, etc. By the same token, we will hardly discuss the commercial applications of optical devices like the telescope or the microscope. In some cases we will indicate that a particular subbranching from the main "E&M tree" will be omitted. Note: The name of the scientists and engineers in this appendix are not included in the general index because this section is already essentially a chronological index.

Key:

* denotes an important discovery or development.

** denotes a very important discovery or development.

*** denotes one of the all-time most important fundamental discoveries or developments.

‡‡- indicates that a subbranch of the "E&M tree" will be henceforth omitted.

Also, when a Nobel Prize is mentioned, it is always in physics, unless otherwise indicated. If the information is available, we will specify the nationality of the scientists mentioned, illustrating the contributions of people from many nations.

Thales of Miletus (640 - 546 BC): The Greek philosopher, is reported to have studied the magnetic properties of lodestone and the electrification of amber. (Note: The Greek word for amber is elektron.)

Ptolemy (100 - 170 AD): The famous Greek astronomer, is reported to have deduced an approximate law for refraction.

Ibn al-Haytham (965-1020): The Muslim physicist, made important advances in experimental optics, using lenses and spherical and parabolic mirrors. He studied spherical aberration, the magnifying power of lenses, and atmospheric refraction. The Latin translation of his work had much influence on Western science, especially through Roger Bacon and Johannes Kepler.

Eleventh century: The Chinese discovered that magnetized needles would align north and south.

*** 1269:** The French scholar Peter Peregrinus observed that like magnetic poles repel and that unlike magnetic poles attract.

1492: The Italian explorer Christopher Columbus discovered the West Indies because he was sailing by his compass, which guided him to the magnetic west. Columbus's voyage had become increasingly difficult and his crew very restless. Thus, it is believed that, if he had been sailing true west, he would not have discovered land soon enough, and his crew would have forced him to turn back to Spain.

*** 1600:** The English physicist William Gilbert demonstrated that a monopole could not be produced by cutting a magnet. He performed extensive magnetic experiments, showing that magnetized iron would become demagnetized under intense heating, but that beating wrought iron can induce magnetism. (Note: The temperature above which a ferromagnet loses its magnetism is now called the Curie point, in recognition of the research done by the nineteenth-century French chemist Pierre Curie.) Also, Gilbert studied declination, the deviation of a magnet from true north, and inclination, or the tendency of a magnet (when swinging in a vertical plane) to point toward the Earth. Thus, he concluded that the Earth was a giant magnet. (Both sides in the geocentric/heliocentric controversy used Gilbert's ideas on magnetism to bolster their points of view; Gilbert himself was a dogmatic advocate of the Copernican theory.) Also, he showed that, suitably rubbed, substances other than amber (electrics) would attract light objects.

1609: The Italian physicist Galileo Galilei is credited with developing the telescope into an important astronomical instrument. (However, the telescope was invented by a Flemish spectacle maker, Hans Lippersley, who received a patent for his device in 1608.) Galileo discovered the moons of Jupiter and the phases of Venus, that the the surface of Earth's moon is rough like that of the Earth's surface, and that the Milky Way consists of many points of light (which he correctly identified as stars). All of these observations convinced him that the Copernican theory was correct. Henceforth astronomy became a precise, empirical science.

‡‡ We will henceforth omit the subbranch dealing with all of inventions and applications of various telescopes and microscopes. ‡‡

1621: The Law of Refraction (Snell's Law) was determined experimentally by Willebrord Snell. The German astronomer Johannes Kepler had shown that Ptolemy's law of refraction was only valid for small angles, but he did not deduce the correct law.

*** ca. 1648:** The French mathematician Pierre de Fermat discovered the Principle of Least Time, proving that the path traversed by light during reflection or refraction satisfies a variational principle. He then deduced for geometric optics the familiar laws of reflection and refraction.

1660: The German Otto von Guericke invented the first electric generator. When a sulfur ball was rotated at high speed by a hand crank, many sparks were produced in a gap connected to two brushes that touched the rotating ball.

* **1665:** The English scientist Sir Isaac Newton used a prism to separate white light into its component colors.

* **1665:** The Italian physicist Mario Grimaldi noted that light passing through two very narrow openings, one behind the other, exhibited slight diffraction. Unfortunately, this very important discovery went unnoticed at the time.

1669: The Danish scholar Erasmus Bartholin discovered double refraction from a transparent crystal known as Iceland spar. This effect could not be explained by either Huygens or Newton, the two main experts on optical phenomena at that time.

** **1676:** The Danish astronomer Ole Roemer determined very approximately the speed of light. This result was obtained by observing Io, one of Jupiter's moons. Roemer noted differences in the measured period of Io, according to whether the Earth was moving toward or away from Jupiter. This important result showed that the speed of light is very large but finite.

** **1678:** Based on his studies of reflection and refraction, Christian Huygens, the Danish physicist, proposed that light was a wave. This challenged the prevailing view that light consisted of particles or "corpuscles". The corpuscular theory was supported by Newton, and because of his great prestige, the wave theory of light was strongly resisted into the next century.

1746: The Dutch physicist Pieter van Musschenbroek at the University of Leyden, invented the Leyden jar, which could store large quantities of electricity.

* **1752:** Using a kite and a Leyden jar, Benjamin Franklin, the American scholar, was credited with demonstrating that lightning is an electrical discharge.

* **1770:** The Swedish chemist Karl William Steele showed that strips of paper soaked in silver nitrate solution darken more rapidly as one proceeds from the red to the violet parts of the visible electromagnetic spectrum.

** **1785:** Benjamin Franklin postulated that there was one type of electricity present in all materials, some of which had an excess of charge (+, positive) when rubbed and others, a deficiency (-, negative). Franklin, however, guessed incorrectly the sign convention for the charges, an error that remains with us today. (Electrical currents are conventionally taken to flow from + to -, whereas we now know that the currents in most conductors are due to negative electrons that move from - to +.)

** **1785:** After inventing the torsion balance, the French physicist Charles-Augustin de Coulomb used it to prove his famous inverse-square force law between two electrical charges. It is also reported that he discovered an analogous law for the strength of a force between magnetic poles. Coulomb believed that electricity and magnetism were "different species of matter whose laws of action were mathematically similar but whose natures were fundamentally different" a belief that would dominate science for the next forty years.

1786: Luigi Galvani, an Italian professor of anatomy, performed many experiments in which amputated frog legs underwent convulsions due to an electrostatic generator. He also showed that the frog legs twitched when connected to lightning rods, thereby demonstrating that lightning is a form of electricity.

* **1800:** The German-British astronomer William Herschel discovered that a thermometer placed at various parts of the visible spectrum recorded successively higher temperatures in proceeding from violet to red. Then, when he placed the thermometer beyond the red end of the visible spectrum, he recorded temperatures that were higher than those in the visible range. Herschel had discovered infrared radiation.

** **1800:** The Italian physicist Alessandro Volta extended Galvani's work (except that Volta used live frogs). He showed that electricity could be produced when two dissimilar metals were immersed in a salt solution. The Voltaic cell consisted of layers of silver and zinc separated by pieces of moist cardboard. Thus, Volta had invented the first electrical battery.

* **1801:** Following Herschel's discovery of infrared radiation, the German chemist Johann Wilhelm Ritter exposed a strip of paper soaked in silver nitrate solution to the part of the visible spectrum beyond the violet. He found that it darkened even more rapidly than a strip exposed using violet light. Thus, Ritter had discovered ultraviolet light.

** **1801:** Thomas Young, the English physicist obtained an interference pattern from his famous double-slit experiment, thereby establishing that light is a wave.

1802: The British chemist William Hyde Wollaston was the first to observe dark (Fraunhofer) lines in the solar spectrum.

* **1814:** The German optician Joseph von Fraunhofer invented the spectroscope, with which he observed hundreds of dark spectral lines (dubbed "Fraunhofer lines") in the Sun's spectrum. He also found that for a white light spectrum the position of the lines was independent of its source.

1814: The British chemists Humphrey Davy and Michael Faraday conducted new experiments involving animal electricity, an ambitious but largely unsuccessful project to which Faraday would return after Davy's death in 1829.

* **1817:** Thomas Young concluded that the double refraction observed by Bartholin in 1669 could be explained if it was assumed that light is a transverse wave.

** **1819:** Augustin-Jean Fresnel reported to the French Academy a transverse wave theory of light that carefully explained intereference, diffraction, and all other known optical phenomena. This settled the nature of light until the next century.

** **1820:** The Danish physicist Hans Christian Oersted showed that a compass needle, when brought near a wire carrying a current, aligned itself at right angles to the direction of current flow. When Oersted reversed the current, the compass pointed in the opposite direction. Thus, an electrical current could influence a magnet. (This was the first strong indication that an electrical current could produce a magnetic field.)

*** **1820:** The French physicist André Marie Ampère showed that if electrical currents flow through two wires in the same direction, the wires attract one another, and if electrical currents flow in opposite directions, the wires repel one another. Ampère clearly understood that the magnetic field forms a circle around a straight current carrying wire. He also showed that a coil of wire could produce a field similar to a bar magnet. His work showed conclusively that a magnetic field was produced by an electrical current. Also, Ampère invented the galvanometer, and from his theoretical work he deduced Coulomb's inverse-square law for magnetism. Jean-Baptiste Biot and Felix Savart, with Ampère, obtained a general formula for the magnetic field arising from a circuit containing a current.

** **1821:** Using iron filings, Michael Faraday demonstrated the existence of magnetic field patterns near various magnets. Faraday was the first scientist to realize that magnetic field lines emanated from the magnet and that one could deal with the physics of fields rather than the object causing the fields. Also, Faraday demonstrated that electricity could produce mechanical action, with magnets made to rotate when placed in current-carrying, mercury-filled bowls. (This was the first, very crude motor.)

** **1827:** The German physicist Georg Simon Ohm established experimentally the famous elementary law bearing his name (V=IR). Ironically, as useful as Ohm's Law has been in the study of electrical circuits, modern microelectronics depends almost entirely on devices for which the law is violated. The law is only approximate, not fundamental. Also, there is a more exact version of the law, which applies to a point inside a homogeneous, isotropic conductor, rather than to a wire.

* **1829:** The American physicist Joseph Henry caused a current to flow through hundreds of turns of insulated wire wound around an iron bar, producing a powerful electromagnet. The reversal of the current flipped the polarity of the magnet. Then, in 1831 Henry used an electrical current to turn a wheel, for which he is credited with inventing the first motor. (Also, apparently Henry discovered electromagnetic induction before Faraday. However, Faraday published his results first and his studies were more extensive than Henry's.)

*** **1831:** Michael Faraday reported on his work, showing (1) that a magnet moving in and out of a wire wound into a helix produces a current in the wire; (2) that an electrical current flows when a copper disc is turned between the poles of a magnet; and (3) that if an iron ring connects primary and secondary coils of wires, a surge of current is induced in the secondary when a current is turned on in the primary. Thus, electricity in wires can be generated by a changing magnetic flux, which is the principle upon which the electric generator is based. At this point in history, there was a fairly complete unification of electricity and magnetism. Faraday also realized that time plays an intimate role in induction effects, an understanding that strongly suggests a velocity is involved in electromagnetic phenomena. Also, Faraday unsuccessfully tried to unify the forces of gravity and electromagnetism.

** **1834:** Heinrich Friedrich Lenz postulated his famous rule, stating that an induced current will tend to flow in a direction opposing the change that produced it (Lenz's Law).

* **1834:** Michael Faraday completed a systematic study of electrolysis. He coined the words *anode, cathode, anion*, and *cation*. He formulated two precise laws of electrolysis.

** **1842:** The classical Doppler effect (or shift) was first proposed by the Austrian physicist Christian Doppler. The effect is well known to us since we experience the following: the pitch of the whistle of a train increases as the train moves toward us and decreases as the train moves away from us. The same behavior applies to light signals, and the modern version of the Doppler shift is obtained from the Special Theory of Relativity (proposed by Einstein in 1905). The Doppler shift has been very important in the spectroscopy component of modern astronomy since the change in frequency of the light spectrum of a distant object allows us to measure its velocity of recession.

** **ca. 1850:** The German physicist Gustav Robert Kirchhoff invented his two laws of circuitry, one of which is based upon the conservation of charge and the other on the conservation of energy. These laws are very important in electrical engineering.

1850: The French physicists Armand Fizeau and Jean-Bernard Foucault made measurements over short distances to determine accurately the velocity of light. They also proved that the velocity of light was less in water than in air, as required by the wave theory of light.

* **1850:** The Italian physicist Macedonio Melloni used prisms made of rock salt to obtain interference fringes from infrared light, thus demonstrating that this radiation exhibits wave properties just like visible light.

** **1859:** The German scientists Robert Bunsen and Gustav Robert Kirchhoff discovered that each element generates its own unique spectrum. They also determined that when hot the element, exhibited a characteristic emission spectrum and when cold, the complementary absorption spectrum. This work ushered in the era of spectral analysis, useful in both chemistry and astronomy.

* **1871:** The British physicist John William Strutt, Lord Rayleigh, demonstrated that the scattering of light exhibits a dependence on the fourth power of the frequency, a result that holds for relatively small frequencies. This Rayleigh scattering explains why the sky is blue and sunsets are red.

1872: The Austrian Ernst Mach asserted Mach's Principle (or Conjecture): In determining the inertial properties of matter, what is important is not absolute acceleration but acceleration with respect to the distant matter of the Universe (the so-called fixed stars). Mach argued that distant matter is responsible for the behavior of Newtonian inertial frames. In layman's terms, "When the subway jerks, it's the fixed stars that throw you down." Some of Machs criticism of Newton apparently influenced Einstein.

*** **1873:** The Scottish physicist James Clerk Maxwell published his famous work (completed between 1864 and 1873) on the four fundamental equations of electromagnetism. These were partial differential equations involving the time and space dependent $\mathbf{E}$ and $\mathbf{B}$ fields. In one of the equations (known as the Ampère-Maxwell Law), he added to Ampère's equation a term involving a changing electrical field (known as "Maxwell's displacement current"), which could produce a magnetic field. Maxwell proved that electromagnetic fields travel at the speed of light, which led him to conclude that light is an electromagnetic wave. (Of course, this analysis also applies to the infrared and ultraviolet parts of the spectrum, which were known at this time.) Thus, electricity, magnetism, and light became unified, which had been a long standing dream of many scientists. Like many other physicists in the 1800s, Maxwell believed in the existence of a mechanical, luminiferous ether to propagate his electromagnetic waves. It has been reported that he eventually abandoned this idea, but unfortunately his work strengthened the belief in the ether theory. Also, Maxwell's original presentation of his work was extremely difficult to comprehend; thus, during the 1870s Oliver Heaviside, an unemployed telegrapher, made the four equations clear and understandable, expressing them in essentially the form that we know today.

* **1878:** The British physicist William Crookes perfected the development of cathode-ray tubes, in which two electrodes were placed in a highly evacuated tube, giving rise to streams of particles passing from the cathode to the anode. Crookes demonstrated that these cathode rays moved in straight lines and could be made to turn a small wheel. (Note: The work on cathode ray tubes began in the 1850s by German scientists, who were intrigued by a strange green glow at the cathode. Crookes's vacuum was much better than that achieved previously.)

* **1879:** The Austrian physicist Joseph Stefan showed that the total radiation emitted by a blackbody increases as the fourth power of the absolute temperature.

** **1884:** The Swedish chemist Svante August Arrhenius completed his thesis study on the dissociation of electrolytes in solution. He argued strongly for the existence of charge-carrying ions, which he insisted had very different properties from their original atoms. Because the atomic theory was not accepted at this time, his thesis committee was very cool to his project (although they reluctantly passed him). He also obtained the general formula giving the rate constant of a chemical reaction. In 1903, Arrhenius won the Nobel Prize in Chemistry for his work.

* **1885:** The Swiss physicist Johann Jakob Balmer discovered his famous empirical formula for the lines of the spectrum of atomic hydrogen.

** **1886:** The German physicist Heinrich Rudolph Hertz used an oscillating circuit with a spark gap and discovered radio waves. He used a rotating mirror to obtain f, the frequency of the waves. Then, with his detecting device (a receiver also containing a spark gap), he could measure the highs and lows of the wave to determine λ, its wavelength. These two measurements gave the velocity of the wave ($v = f\lambda$), which was the speed of light. These waves, which are part of the electromagnetic spectrum, had wavelengths much longer than any previously observed radiation.

*** **1887:** The famous experiment performed by the Americans Alfred Michelson and Edward Morley demonstrated that the velocity of the Earth through the ether could *not* be detected. This result challenged the theoretical foundations of classical physics and eventually ushered in Einstein's Special Relativity..

* **1890s:** The German physicist Wilhelm Wien studied experimentally the spectrum from blackbody radiation. He found that the very long and very short wavelengths are suppressed and that the spectrum has a peak, which is shifted to shorter wavelengths as the temperature increases. Also, he worked out a formula that took into account the high-frequency limit, but not the low-frequency limit, of the spectrum. (About the same time Lord Rayleigh and Sir James Jeans developed a formula for explaining the low-frequency, but not the high-frequency, limit.) Wien received the Nobel Prize for his work in 1911.

* **1892:** Hertz's student Philipp Eduard Anton von Lenard showed that cathode rays would penetrate through a thin aluminum window in the tube and penetrate about eight centimeters into the surrounding air. He concluded that the rays were not electromagnetic radiation. Instead they were particles smaller than air molecules, which he called "quanta of electricity". Von Lenard won the Nobel Prize in 1905.

** **1895:** Studying luminescence of chemicals with cathode rays, the German physicist Wilhelm Konrad Röntgen discovered x-rays, for which he received the Nobel Prize in 1901. At this time it was not known whether x-rays were streams of particles or some type of electromagnetic radiation. Today x-rays are very useful in medicine.

* **1896:** The Dutch physicist Pieter Zeeman showed that the spectral lines produced by sodium are split by a magnetic field (the Zeeman Effect), for which he received the Nobel Prize in 1902.

* **1896:** The French physicist Antoine Henri Becquerel discovered radioactivity while studying the fluorescence of a compound containing uranium. Also, he suspected that more than one type of radiation was involved. Becquerel won the Nobel Prize in 1903.

** **1897:** By allowing cathode rays to speed between two charged metal plates, the British physicist Joseph John Thomson detected a distinct curvature of a negatively charged particle. Thus, cathode rays consist of small, negatively charged particles. Thomson had discovered the electron, the carrier of the current in wires and the first of the modern elementary particles. Thomson also used crossed electric and magnetic fields to measure the mass-to-charge ratio of the electron. He received the Nobel Prize in 1906 for this work. Also, his methods govern the operation of the tubes found in older television sets.

* **1898:** The Polish-French physicist Marie Curie determined that thorium also exhibited radioactivity. Like Becquerel, Madam Curie also suspected several types of radiation.

Madam Curie won the Nobel Prize in 1903. Then, in 1911, she won the Nobel Prize in Chemistry. Remarkably, she was the only person to win Nobel prizes in two different fields.

** **1899:** The New Zealander-British physicist Ernest Rutherford discovered two of the radioactive emanations: alpha particles which could be stopped by 1/500 of a centimeter of aluminum, and beta rays, which penetrated more deeply. Both of these particles could be deflected by a magnetic field. Very soon the beta particle was determined to be a negatively charged electron, and eventually Rutherford (1909) showed that the alpha particle is a positively charged helium nucleus. For such contributions Rutherford received the Nobel Prize in Chemistry in 1908.

** **1900:** The French physicist Paul Ulrich Villard discovered the third, most deeply penetrating radioactive emanation: gamma rays, which were not deflected by a magnetic field. Rutherford had missed this radiation because it had passed completely through his apparatus. It was reported that Rutherford later passed the gamma rays through a crystal and witnessed a diffraction pattern similar to that of x-rays, with the gamma rays possessing a much shorter wavelength than that of x-rays. (Presumably this work could not have been done until after the work on x-rays by Max von Laue in 1912 and William Lawrence and William Henry Bragg in 1913.) Thus, gamma rays are the most energetic part of the electromagnetic spectrum. This radiation arises from transitions between nuclear energy levels.

*** **1900:** The German physicist Max Karl Ernst Ludwig Planck discovered the correct formula for the blackbody spectrum, by assuming that the radiation is emitted in discrete quanta. This was the beginning of the quantum theory. In 1918 Planck won the Nobel Prize.

1901: The Italian Guglielmo Marconi invented and developed the wireless telegraph, with a message being sent across the Atlantic Ocean in Morse code. This was the beginning of long-distance radio communication. For their contributions to wireless telegraphy, Marconi and Karl Ferdinand Braun shared the Nobel Prize in 1909.

‡‡ We will henceforth omit the subbranch dealing with the many practical inventions involving electrons and electromagnetic waves such as the telephone, electrical lights, radio, television, electronics, computers, x-ray devices, etc. However, we would like to acknowledge the American Thomas Alva Edison, the Austrian Nikola Tesla, and the Prussian Charles Steinmetz, all inventors who in the late 1900s made very important contributions to commercial applications of E&M, mainly in electric power generation, but also in telegraphy, X-ray imaging, and motion pictures. ‡‡

* **1902:** Philipp von Lenard studied the photoelectric effect (which had been known since the 1880s) and showed that the speed of the electrons ejected depends upon the frequency of the incident light (if the frequency exceeds a certain "threshold value"), but the number of electrons depends upon the intensity of the light. It was also established that different metals have different threshold values. These experiments could not be explained by nineteenth-century physics.

*** **1905:** The German-Swiss mathematician/physicist Albert Einstein published a famous paper showing that the photoelectric effect is due to the type of quantum behavior first proposed by Planck. This very important result established that light has a particle-like behavior. Einstein determined that the energy of the particle (a photon), increases with its frequency. Thus, light is both a particle and a wave (a "wavicle", if you like). This "particle-wave duality" dictates that whether you observe light as a wave or as a particle depends upon the nature of your experiment. For this work, Einstein received in 1921 his only Nobel Prize. Also, ironically this work established Einstein as one of the founders of Quantum Mechanics, a theory he later came to reject because it is not deterministic. (In his famous letter to Max Born, he said that "You believe in a God who plays dice, and I in complete law and order.")

*** **1905:** Albert Einstein published his work on the Special Theory of Relativity, which assumed that the speed of light in a vacuum is a constant independent of the frame of reference. Moreover, the theory predicts that the physical (group) velocity of any particle can never exceed the speed of light. This theory completely explained the puzzle of the Michelson-Morley experiment. Also, it turned out that relativity allowed the comparatively recent equations of Maxwell to remain intact, while Newton's Second Law (dating back more than two hundred years) had to be modified. Also, in relativity the famous equation $E = mc^2$ was derived, a result published in a separate paper in 1905. Incidentally, in 1905 there were two other very important papers published by Einstein: his dissertation, detailing how to measure the size of molecules and Avogadro's number, and a fundamental companion paper on Brownian motion. Never before or since in the history of science have we witnessed such exceptional creativity in a very short period of time (with the possible exception of Newton)! And Einstein completed all of this work in his spare time while working as a clerk in the Swiss patent office!

** **1909:** The development of modern atomic and nuclear theory began with the famous experiment of Ernest Rutherford and Ernest Marsden, in which alpha particles were scattered through a thin film of gold. It was a great surprise to learn that occasionally an alpha particle was deflected directly backward from its original direction. It was evident that exceptionally strong forces are present in a tiny region (the nucleus, $10^{-13} - 10^{-12}$ cm), which is many orders of magnitude smaller than the atom (10^{-8} cm). See appendix SP.1 on Brief Survey of Developments in Atomic, Molecular, and Nuclear Physics, in which we also summarize the relevance to the history of E&M.

‡‡ Henceforth, we will omit the subbranch dealing with atomic, molecular, and nuclear structure, but we will include certain fundamental work associated with elementary particles and electromagnetic phenomena. ‡‡

** **1911:** The Danish physicist Heike Kamerlingh Onnes discovered superconductivity. He found that, in a ring of mercury cooled to temperatures below 4.2 K, a steady current would flow without being sustained by an electrical field, from which he concluded that the electrical resistivity of certain metals and other materials vanish if cooled to sufficiently low temperatures. This work directly followed from the experimental studies in 1908 in which Onnes learned how to liquefy helium, thereby making temperatures down to about 1 K available. In 1913 Onnes won the Noble Prize.

** **1911:** The American Robert Millikan measured the charge on the electron in his famous oil-drop experiment. Millikan found that the measured charges on the oil droplets were always integer multiples of a single number, the charge of the electron. This result, coupled with that of J. J. Thomson's experiment, allowed the mass of the electron to be accurately determined. Millikan won the Nobel Prize in 1923 for this work.

* **1912:** The German physicist Max Theodor Felix von Laue showed that x-rays could be diffracted by a crystal of zinc sulfide. The diffraction pattern was that expected of electromagnetic radiation of very short wavelength. In 1914 von Laue won the Nobel Prize for this work. x-rays arise from transitions involving the innermost energy levels of atoms.

** **1912:** The American astronomer Henrietta Leavitt discovered Cepheid-variable stars in the Small Magellanic Cloud satellite galaxy. Leavitt showed that the period of a Cepheid is related in a linear fashion to its intrinsic brightness, a property allowing astronomers to measure the distance to the Cepheids.

* **1913:** The British father and son team William Henry Bragg and William Lawrence Bragg showed that the precise wavelength of diffracted x-rays could be determined if the crystal lattice spacing were accurately known. For their x-ray diffraction studies of crystals, these two physicists shared the Nobel Prize in 1915.

*** **1916:** Albert Einstein published his General Theory of Relativity, which deals with the curvature of the SpaceTime continuum and explains gravitational forces. One of the central postulates of the General Theory is the "The Principle of Equivalence", which states that the effects of an acceleration are indistinguishable from those of a gravitational field. There are two parts of this theory that are especially applicable to E&M: The first is gravitational lensing. Starlight passing near a star is deflected twice as much as in Newtonian theory. Such an effect was first observed by Sir Arthur Eddington in a total eclipse of the sun in 1919. Another example is the light emitted by distant quasars, which is bent by intervening galaxies. The second part is the gravitational red shift which is the slowing down of time (or the lowering of frequencies) as the gravitational field is increased. Such an effect has been observed in the light emitted from white dwarf stars. Also, this red shift would cause any object moving toward a Black Hole to move ever more slowly as it encounters more and more the intense gravitational field there. By the same token, nothing can escape from a Black Hole, not even light (except for the emission from mini Black Holes of Hawking radiation, which has never been observed). The General Theory clearly predicts Black Holes, an example of which may be an invisible object that is believed to orbit a luminous blue giant star in the Cygnus constellation. (Another example is a quasar whose bright, but very red-shifted, light is thought to arise from disintegrating stars, which are streaming toward a giant black hole in second-generation star formation of an early galaxy.) Also, the General Theory of Relativity is the theoretical basis for the Big Bang cosmology.

1923: Although the photon was implied by Einstein's work on the photoelectric effect in 1905, its official discovery is credited to the American physicist Arthur Holly Compton who scattered x-rays from atomic electrons in a crystal spectrometer. Compton also apparently named the photon and showed that the shorter the wavelength, the more the photon behaves like a particle.

** **1925:** The Dutch physicists Samuel Goudsmit and George Uhlenbeck proposed that the electron possesses an intrinsic "spin", an explanation that removed various difficulties with the spectral lines in the Bohr theory of the atom. In particular, since this spin produces an intrinsic magnetic moment of the electron, the nature of the Zeeman effect was understood—namely, the splitting of the spectra of atoms when subjected to magnetic fields. (There is also a contribution to the Zeeman splitting due to the magnetic moment arising from the orbital motion of the electron.) It should be mentioned too that Wolfgang Pauli had postulated for the electron the existence of another quantum number, which was essentially intrinsic spin.

*** **1927:** The theory of intrinsic spin was put on a firm theoretical foundation by the British physicist Paul Dirac in his famous relativistic wave equation for the electron, for which he earned the Nobel Prize in 1933. Dirac's theory also established the existence of antiparticles; in particular, Dirac predicted the positron, a particle with the same mass and spin but the opposite charge of the electron. In 1932 the American physicist Carl David Anderson discovered experimentally the positron, for which he was awarded the Nobel Prize in 1936. Later, in 1955, Emilio Segre and collaborators discovered the antiproton and in 1956 Bruce Cork and collaborators discovered the antineutron.

‡‡ We will henceforth omit most of the subbranch dealing with all of the many important and in some cases very short-lived particles and antiparticles that have been discovered in this century. Most of these particles are charged and therefore experience the electromagnetic force. ‡‡

1929: The American scientist and inventor Edwin Land invented the first inexpensive polarizing filter, Polaroid film.

***** 1929:** The American astronomer Edwin Hubble announced his famous result that the recession speed of a galaxy is (roughly) linearly proportional to its distance from the Earth, the constant of proportionality being the Hubble constant the inverse of which gives the age of the Universe. The recession speed is obtained by analyzing the relativistic Doppler shift for the frequency of light from stars in the galaxy. This work showed that Universe is expanding.

*** 1930:** The Italian Enrico Fermi studied the hyperfine splitting of the ground-state energy level of atomic hydrogen, which is due to the interaction of the magnetic moments of the electron and the proton. Electronic transitions between these two levels produces the famous 21 cm wavelength radiation observed in radio astronomy, which gives scientists much information about the motion and distribution of hydrogen gas in interstellar space. Fermi also perfected an early theory of the weak interactions.

1931: In normal E&M theory, unlike an electric charge, a monopole cannot be isolated. Thus there is an asymmetry in Maxwell's equations between electric and magnetic forces, which has led various theorists to speculate about the existence of magnetic monopoles. In this year Dirac developed a symmmetric version of Maxwell's equations in order to explain the quantization of electric charge.

**** 1933:** The German physicists Walther Meissner and Robert Ochsenfeld discovered that in the superconducting state a material becomes a "perfect diamagnetic," with magnetic flux completely expelled from the interior of the sample, (the Meissner effect.)

1934: The German brothers Fritz and Heinz London developed a simple series of classical equations that approximated the Meissner effect in superconductors.

***** 1940–1950:** The deluxe theory of electromagnetism finally arrived—Quantum Electrodynamics (QED), which was worked out in great detail by many physicists. However, the most important contributions were made by the Americans Richard Feynman and Julian Schwinger, and the Japanese Sin-Itiro Tomonaga, for which they shared the Nobel Prize in 1965. In QED classical electromagnetism is combined with both Quantum Mechanics and Special Relativity, and the theory describes the quantum interactions between charged particles and the electromagnetic field.

*** 1950:** The Russian physicists Vitaly Ginzberg and Lev Landau developed a highly successful phenomenological theory of superconductivity that replaced the simpler London equations.

**** 1952:** The German astronomer Walter Baade, who worked in the United States from 1893 to 1960, announced that he had discovered an error in the previous distance calculations of Cepheid variable stars. Baade had discovered that there are two types of Cepheids, which resulted in both a doubling of the size of the known Universe and the age of the Universe.

1956: The American physicists William Shockley, John Bardeen, and Walter Houser Brattain won the Nobel Prize for their research on semiconductors and the discovery of the transistor.

***** 1956:** The Chinese-American physicists Chen Ning Yang and Tsung-Dao Lee predicted that reflection symmetry (P, or parity) is violated in the weak interactions, a result that was soon confirmed experimentally by the Chinese-American physicist Chien Shiung Wu. Yang and Lee won the Nobel Prize for their work in 1957. Very soon thereafter it was established that charge conjugation symmetry (C) is also violated in the weak interactions. Under C, the charge of a particle is reversed (as well as other quantum numbers such as lepton number), so that a particle and its antiparticle are interchanged. Then, in 1964 the American physicists Val Logsdon Fitch and James Watson Cronin proved that CP symmetry (the product of C and P) is violated in the weak interactions, for which they won the Nobel Prize in 1980.

**** 1957:** The Americans John Bardeen, Leon N. Cooper, and J. Robert Screiffer developed a quantum mechanical theory of superconductivity (the BCS theory), in which a highly collective quantum state is occupied by electrons bound in pairs. Because their theoretical predictions were in excellent agreement with experiment, they won the Nobel Prize in 1972. (It should be noted too that John Bardeen is the only scientist to have earned two Nobel Prizes in the same field.) Also, in the 1960s BCS Theory was successfully applied to nuclear physics in order to explain pairing effects.

*** 1958:** The Russian physicist Alexei Abrikosov introduced the concept of a superconducting vortex, a discovery for which he won the Nobel Prize in 2003.

***** 1961–1968:** A unified theory of the weak and electromagnetic interactions was developed by the Americans Steven Weinberg and Sheldon Glashow and the Pakistani physicist Abdus Salam (the Glashow-Weinberg-Salam ElectroWeak Theory.) In this theory, the weak and electromagnetic forces are unified at a high temperature or energy (such as that found in the early universe or in a particle accelerator), but the "symmetry is broken" for our present low-energy existence. Glashow, Weinberg, and Salam received the Nobel Prize for this work in 1979. The ElectroWeak theory, when combined with the modern theory of the strong interactions (QCD) is known as the Standard Model, whose validity is now well established experimentally.

***** 1963:** Using $SU(3)$ group theory, the American Murray Gell-Mann and the Israeli Yuvall Neeman proposed the existence of small particles called quarks that possess fractional electrical charge. In particular, the magnitudes of the charges of the quarks are in units of $1/3$ or $2/3$ of the magnitude of the charge on the electron. For the first time, it became clear that electrical charge was not quantized in integer units of the electron charge. The experimental discovery of the Ω^- particle in 1964 confirmed Gell-Mann's theory, and he was awarded the Nobel Prize in 1969. The quarks are now known to be the constituent particles making up neutrons, protons, and other particles (known collectively as hadrons).

***** 1964:** The German-American Arno Allan Penzias and the American Robert Woodrow Wilson, both radio astronomers, discovered the 2.7 K cosmic microwave background radiation, which is thought to originate about 300,000 years after the Big Bang (when radiation decoupled from matter). This important result is one of the cornerstones of the Big Bang Theory, and Penzias and Wilson won the Nobel Prize in 1978. Also, this discovery answered a very important question, perhaps first proposed by Marconi: In SpaceTime what are the limits of communication with electromagnetic waves? The answer is that we can see radiation from almost the beginning of the universe.

1967: Baryogenesis. In order to explain the almost complete dominance of matter over anti-matter in the universe, the Russian physicist Andrei Sakharov advanced a theory, involving both the violation of baryon conservation and the violation of C and CP symmetries in the weak interactions.

1968: The British astronomer Gerrit L. Verschuur discovered the first evidence for the Zeeman effect from the splitting of the 21 cm radio wavelength due to hydrogen in interstellar gas clouds.

1975: Richard Blakemore, a microbiologist from the Woods Hole Oceanographic Institution discovered magnetotactic bacteria that swim along the Earth's magnetic field lines.

1982: On Valentines Day, using a coil of superconducting niobium and a superconducting magnetic shield, Blas Cabrera at Stanford received a signal indicating the possible passage of a magnetic monopole, the only such event that he has recorded. In 1986 at Imperial College in London, another possible monopole signal was seen. Both of these events are considered very questionable.* However, the search continues because of the interest of high-energy physicists in grand unified theories (GUTs, which involve the unification of the strong and ElectroWeak forces), some of which predict the existence of magnetic monopoles with enormous masses ($\approx 10^{16}$ GeV), near the unification energy. Thus, magnetic monopoles could have been produced in the early universe. One explanation for the apparent absence of monopoles involves Alan Guth's inflationary-universe hypothesis (1979). Inflation allows for a very rapid expansion of the Universe ($\approx 10^{-35} - 10^{-33}$ sec after the Big Bang), which eliminated large mismatches in energy fields that could have contributed to magnetic monopole production.

*** 1987:** The Swiss physicist Karl Alex Muller and the German physicist Johannes George Bednorz won the Nobel Prize for the discovery of a new class of "high-T_c" superconductors. In these substances, often complex compounds, the critical temperature exceeds the boiling point of liquid nitrogen, thereby allowing the superconducting state to be achieved at relatively high temperatures. It is hoped that critical temperatures around room temperature may eventually be achieved.

**** 1987:** The Americans Derek Lovely, John Stolz, Gordon Nord, Jr., and Elizabeth Phillips have established that a bacterium GS-15 (an anaerobic organism that requires iron for its metabolism) converts ferric oxide to lodestone particles. This is likely the mechanism responsible for the formation of the lodestone existing in the Earth and without which we would have learned about magnetism much more slowly. (This, in turn, could have greatly slowed down the developments of most of our present technology.)

1989: The Cosmic Background Explorer (COBE) was launched, the purpose of which was to obtain much more detailed measurements of the cosmic background radiation. A central purpose of the project was to measure accurately the fluctuations in the radiation since these could explain the large- scale structure of galaxies and supergalaxies that we observe today in the universe.

---

*Exactly a year after the original event, Sheldon Glashow (one of the founders of the ElectroWeak Theory and a closet poet) sent Cabrera the following Valentine's message:

Roses are red,
violets are blue,
The time has come
for Monopole II.

* **1991:** From measurements of rockets and satellites, scientists were able to show that the cosmic background radiation exhibited very closely a blackbody spectrum of ≈ 3 K. Then, a precision instrument aboard COBE showed that the precise value of the blackbody temperature is 2.726 ± 0.005 K.

*** **1992:** Fluctuations, or anisotropies, in the cosmic microwave background radiation ($\approx$ 1 part in 100,000) were first seen with COBE. This work proved conclusively how galaxies formed in the very early Universe and established once and for all the Big Bang cosmology. This work was led by the Americans John Mather and George Smoot, who won the 2006 Nobel Prize.

*** **1998:** Two groups used ground and space telescopes to examine distant 1a supernovae, whose light seemed dimmer than expected. It was found that the data meant that the Universe is not only expanding, but it is accelerating. For one group that investigated the so-called "high-z, type Ia supernova" data, a good fit is given by: for dark energy density, called $\Omega_\Lambda, \approx 0.7$ and for total matter density, $\Omega_M, \approx 0.3$,[†] where Ω is the ratio of a particular density to the critical density required to close the Universe. The researchers for this work point out too that $\Omega_M \approx 0.3$ and $\Omega_\Lambda \approx 0.7$ arise too from combining high-z, supernova data (which measures $\Omega_M - \Omega_\Lambda$) with data from fluctuations in the cosmic microwave background radiation (which measures $\Omega_M + \Omega_\Lambda$). Note that matter is responsible for gravitational attraction and tends to decelerate the Universe, while dark energy gives a kind of negative pressure (anti-gravity), tending to accelerate the Universe. The case has been made that inflationary theory, which is dear to the hearts of many physicists, is essentially saved by the new theory, which establishes that $\Omega = \Omega_M + \Omega_\Lambda = 1.0$. For this important work, the 2011 Nobel Prize was shared by three astrophysicists: the Americans Saul Perlmutter and Adam Riess and the Australian Brian Schmidt.

Appendix SP.1. Brief Survey of Developments in Atomic, Molecular, and Nuclear Physics

Atomic and Molecular Physics. For modern atomic and molecular structure, two important points pertaining to E&M history should be emphasized:

(1) Emission and absorption spectra from atoms and molecules can be understood quantitatively.

(2) Atoms and molecules are held together by the electromagnetic force.

** In 1911 Rutherford advanced his idea about atomic structure. Then, in 1913 the Danish physicist Niels Bohr proposed his model of the atom ("the old Bohr theory"), for which he received the Nobel Prize in 1922. Bohr's theory gave a semiclassical explanation for the spectral lines of the hydrogen atom. Other important contributions were made in 1916 by the German physicist Arnold Johannes Wilhelm Sommerfeld and by the Austrian-Swiss physicist Wolfgang Pauli, who in 1925 formulated the famous Pauli Exclusion Principle (for which he won the Nobel Prize in 1945).

[†]It has also been determined that the enigmatic dark matter is the dominant form of matter in the Universe. It is found that Ω(baryonic matter) ≈ 0.04 and Ω(baryonic matter resulting in visible light) ≈ 0.005. Thus, assuming that the Universe has a total mass/energy density $\Omega = \Omega_M + \Omega_\Lambda = 1$, less than 5% of this total density is baryonic matter (with less than 1% of the total due to visible baryons) and the total matter or mass density is only about 30% (the great majority of which is non-baryonic dark matter, which could be neutrinos, axions, or super-symmetric particles). Then, the rest of Ω ($\approx 70\%$) is made up of dark energy.

*** Quantum Mechanics was then used to explain atomic and molecular structure. First, in 1924 the French physicist Louis Victor de Broglie proposed that particles exhibit a wavelike nature, for which de Broglie received the 1929 Nobel Prize. This idea was confirmed in 1927 by the American physicists Clinton Davisson and Lestor Germer and independently by the British physicist Geord Paget Thomson (both groups by using diffraction through a crystal); Davisson and Thomson shared the Nobel Prize in 1937. Next, atomic structure was put on a firm theoretical basis by the discoveries of (1) matrix mechanics in 1925 and the uncertainty principle in 1927, both by the German physicist Werner Karl Heisenberg (Nobel Prize, 1932) and (2) wave mechanics in 1926 by the German physicist Erwin Schrödinger (Nobel Prize, 1932). Finally, in 1939 quantum mechanical studies of the transference and sharing of electrons between atoms, were completed by the American chemist Linus Carl Pauling (Nobel Prize in Chemistry, 1954).

*** As mentioned previously, in 1927 Paul Dirac formulated relativistic Quantum Mechanics, which combined special relativity and Quantum Mechanics and which predicted an intrinsic spin of the electron. Dirac received the Nobel Prize in 1933.

Nuclear Physics. The main relevance to E&M is that at the time of the 1909 Rutherford-Marsden experiment, it was not clear what mechanism prevents the positively charged nucleus from being blown apart due to the intense Coulomb repulsion. The answer, of course, is that among the neutrons and protons that make up the nucleus, there is a strong, attractive nuclear force that counteracts the repulsive force. Also, it was found that the nuclear force is charge independent (i.e., the nuclear interaction between a neutron and a proton is the same as that between a proton and a proton, or a neutron and a neutron). Thus, as we move up the periodic table to heavier nuclei that contain more protons, neutrons must be added in order to stabilize the nucleus. Also, once nuclear energy levels were calculated and measured, nuclear physicists were able to determine the energies of the gamma rays emitted or absorbed by transitions between the levels.

** The proton was discovered by Rutherford's team from 1911 to 1919. Then, in 1932 the British physicist James Chadwick detected the neutron in an ionization chamber when bombarding beryllium with alpha particles. For this work, Chadwick received a Nobel Prize in 1935. Many more important discoveries were made over the next fifty years, culminating in the Nobel Prizes won in 1963 by Maria Goeppert Mayer and J. Hans D. Jensen (for studies of nuclear shell structure) and in 1975 by Aage Niels Bohr, Ben Roy Mottelson, and Leo James Rainwater (for studies of the connection between particle and collective motion in the nucleus).

The Galilean and Lorentz Equations;
The Invariance of Proper Time

A.1. The Galilean Transformation

See Fig. A.1. The classical (i.e.: pre-relativity) Galilean equations are given by

$$x' = x - ut \tag{A.1a}$$

$$y' = y \tag{A.1b}$$

$$z' = z \tag{A.1c}$$

$$t' = t, \tag{A.1d}$$

which are very intuitive. In particular Eq. (A.1a) is easily seen from the construction in Fig. A.1. Notice that the perpendicular coordinates do not change. Also, Eq. (A.1d) implies that in pre-relativistic physics, there is an absolute time (this was strongly believed by Newton and others).

The inverse transformation from (x', y', z', t') to (x, y, z, t) is easily obtained from Eqs. (A.1):

$$x = x' + ut \tag{A.2a}$$

$$y = y' \tag{A.2b}$$

$$z = z' \tag{A.2c}$$

$$t = t'. \tag{A.2d}$$

These can be obtained algebraically from Eqs. (A.1), or by letting $u \to -u$ in Eq. (A.1a). Obviously if S' moves with a frame velocity u with respect to S, then S moves with a frame velocity $-u$ with respect to S'.

A.2. The Lorentz Transformation

Here we follow the derivation in Ref. 2, pages 33 - 38.

We again use the the configuration of the frames shown in Fig. A.1. We make three important assumptions:

(i) In the transformation, the x' and t' coordinates depend linearly on the x and t coordinates, so that space and time get mixed up. In particular, $t' \neq t$.

(ii) We demand that, in the limit of small values of u, we reproduce the Galilean equations (A.1). Also, note that once again, the perpendicular coordinates are unchanged; ie:

$$y' = y \tag{A.3a}$$
$$z' = z. \tag{A.3b}$$

(iii) The velocity of light is the same in each inertial frame. Moreover, the magnitude of the velocity of light is the constant c in each inertial frame.

Note that in assumption (i) we say that the transformation equations should be linear (in analogy with the Galilean transformation). That means that there can only be four coefficients relating x and t to x' and t', one of which is simply eliminated by assumption (ii) (since the final combination must depend on $x - ut$).

Then, from assumptions (i) and (ii), we obtain the transformation equations:

$$x' = \gamma(u)(x - ut) \tag{A.4a}$$
$$y' = y \tag{A.4b}$$
$$z' = z \tag{A.4c}$$
$$t' = \alpha(u)t + \beta(u)x, \tag{A.4d}$$

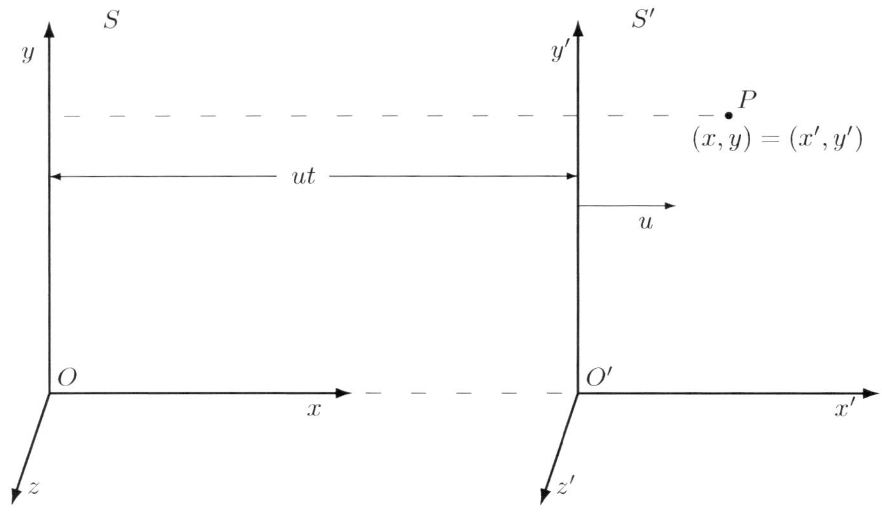

FIGURE A.1. The inertial frame S' $[(x', y', z')]$ moves with a positive frame velocity u along the positive x and x' axes with respect to the stationary intertial frame S $[(x, y, z)]$. It is assumed that at $t = t' = 0$ the two origins coincide, with $x = x' = y = y' = z = z' = 0$.

and we note that the transformation coefficients α, β, and γ can depend on the frame velocity u. Also, from assumption (ii) we demand that

$$\alpha(u) \to 1 \quad \text{as} \quad u \to 0 \tag{A.5a}$$

$$\beta(u) \to 0 \quad \text{as} \quad u \to 0 \tag{A.5b}$$

$$\gamma(u) \to 1 \quad \text{as} \quad u \to 0. \tag{A.5c}$$

Next, at $t = t' = 0$ when the origins of the two systems coincide ($\vec{r} = \vec{r}' = 0$), we assume the emission of a light signal. At later times, the envelope of the light signal is given by:

$$r^2 = x^2 + y^2 + z^2 = c^2 t^2 \tag{A.6a}$$

in S, and

$$r'^2 = x'^2 + y'^2 + z'^2 = c'^2 t'^2 \tag{A.6b}$$

in S', where c and c' are velocities of light in S and S' respectively. However, from assumption (iii)

$$c = c', \tag{A.7}$$

and from Eqs. (A.4b) and (A.4c)

$$y^2 + z^2 = y'^2 + z'^2, \tag{A.8}$$

which implies

$$c^2 t^2 - x^2 = c^2 t'^2 - x'^2. \tag{A.9}$$

Note too that we focus on a particular point on the envelope at which $y = y', z = z'$. We use Eqs. (A.4a), (A.4d) and (A.9) to determine the relationships between x, t and x', t'.

Substituting Eqs. (A.4a) and (A.4d) into Eq. (A.9) and matching coefficients of t^2, x^2 and tx, we obtain

$$c^2 \alpha^2 - \gamma^2 u^2 = c^2 \tag{A.10a}$$

$$c^2 \beta^2 - \gamma^2 = -1 \tag{A.10b}$$

$$2\alpha\beta c^2 + 2\gamma^2 u = 0. \tag{A.10c}$$

After some manipulations, we find that

$$\gamma^2 = \left(1 - \frac{u^2}{c^2}\right)^{-1} \tag{A.11a}$$

$$\alpha^2 = \frac{c^2}{(c^2 - u^2)} \tag{A.11b}$$

$$\alpha\beta c^2 = -\gamma^2 u. \tag{A.11c}$$

From conditions (A.5a) and (A.5c) we have

$$\gamma = + \left(1 - \frac{u^2}{c^2}\right)^{-\frac{1}{2}} \tag{A.12a}$$

$$\alpha = + \left(1 - \frac{u^2}{c^2}\right)^{-\frac{1}{2}} = \gamma \tag{A.12b}$$

$$\beta = -\frac{\gamma u}{c^2}. \tag{A.12c}$$

We see too that condition (A.5b) is satisfied. Thus, we obtain the Lorentz transformation:

$$x' = \gamma(x - ut) \tag{A.13a}$$

$$y' = y \tag{A.13b}$$

$$z' = z \tag{A.13c}$$

$$t' = \gamma\left[t - \frac{u}{c^2}x\right], \tag{A.13d}$$

with

$$\gamma = \left(1 - \frac{u^2}{c^2}\right)^{-\frac{1}{2}}. \tag{A.12a}$$

Note too that, in the derivation,

$$|u| < c$$

in order that α, β, and γ be real and finite. This restriction was not assumed, but it comes out of the transformation derivation.

By letting $u \to -u$ in Eqs. (A.13) we obtain the inverse equations:

$$x = \gamma\left(x' + ut'\right) \tag{A.14a}$$

$$y = y' \tag{A.14b}$$

$$z = z' \tag{A.14c}$$

$$t = \gamma\left[t' + \frac{u}{c^2}x'\right]. \tag{A.14d}$$

Note that $\gamma(-u) = \gamma(u)$.

We can also define the four-vector

$$X_\mu = (\vec{r}, ct) = (x, y, z, ct)$$
$$= (\vec{r}, X_t), \tag{A.15}$$

from which we can rewrite Eqs. (A.13)

$$x' = \gamma\left(x - \frac{u}{c}X_t\right) \tag{A.16a}$$

$$y' = y \tag{A.16b}$$

$$z' = z \tag{A.16c}$$

$$X_t' = \gamma\left(X_t - \frac{u}{c}x\right). \tag{A.16d}$$

In this form, we see the symmetry between Eqs. (A.16a) and (A.16d).

Finally, it is possible to write down a general Lorentz transformation between two frames, in which one of the frames S' moves with a general velocity $\vec{u}$ with respect to the other (S) frame. However, these equations are complicated and unnecessary since one can always imagine transforming (translations and rotations) the two frames so that they coincide at $t = t' = 0$ and $\vec{r} = \vec{r}' = 0$, with the velocity $\vec{u}$ along the x and x' axes.

A.3. The Invariance of Proper Time Under a Lorentz Transformation

We define the differential of proper time τ as

$$c^2 \mathrm{d}\tau^2 = c^2 \mathrm{d}t^2 - \mathrm{d}r^2, \tag{A.17}$$

where

$$\mathrm{d}r^2 = \mathrm{d}x^2 + \mathrm{d}y^2 + \mathrm{d}z^2. \tag{A.18}$$

From Eqs. (A.13) and (A.17), it can easily be shown that

$$c^2 \mathrm{d}\tau^2 = c^2 \mathrm{d}\tau'^2, \tag{A.19}$$

so that the proper time is invariant under a Lorentz transformation.

In fact, the differential of the proper time can be written in terms of the four-vector given by Eq. (A.15):

$$\mathrm{d}X_\mu = (\mathrm{d}\vec{r}, c\,\mathrm{d}t), \tag{A.20}$$

from which we obtain the four-dimensional dot product:

$$\mathrm{d}X_\mu \cdot \mathrm{d}X_\mu = \mathrm{d}r^2 - c^2 \mathrm{d}t^2 = -c^2 \mathrm{d}\tau. \tag{A.21}$$

Then, as shown in the main text (in Problem #1, Chapter 4), the four-dimensional dot product of a four-vector is an invariant under a Lorentz transformation. This again establishes Eq. (A.19).

Indeed, an alternative derivation of the Lorentz equations is to start with the differentials of Eq. (A.4)

$$\mathrm{d}x' = \gamma(\mathrm{d}x - u\,\mathrm{d}t) \tag{A.22a}$$

$$\mathrm{d}y' = \mathrm{d}y \tag{A.22b}$$

$$\mathrm{d}z' = \mathrm{d}z \tag{A.22c}$$

$$\mathrm{d}t' = \alpha\,\mathrm{d}t + \beta\,\mathrm{d}x \tag{A.22d}$$

and demand that Eq. (A.19) be satisfied. By the same procedure outlined in subsection A.2, we again find Eqs. (A.13).

Thus, a necessary and sufficient condition for the validity of the Lorentz equations is that the proper time be an invariant.

Appendix **B**

The Photon Clock: an Alternative Derivation of the Lorentz Contraction

Attach two mirrors at the two ends of a ruler (in S'), and let the lower end contain a flashlight or other device for emitting photons. This arrangement gives a *Photon Clock*.

The ruler moves with a velocity u with respect to a stationary frame S as in Fig. B.1. At $t = 0$, a photon is emitted from the back mirror. The photon strikes the front mirror at (t_2, x_2), after which it is reflected. The photon then strikes the back mirror at (t_1, x_1).

From the geometry in Fig. B.1 and using Eqs. (A.14a) and (A.14d), we find that

$$l = x_2 - x_1 + u(t_1 - t_2) = \frac{x_2' - x_1' + u(t_2' - t_1')}{\sqrt{1 - \frac{u^2}{c^2}}} + \frac{u\left[(t_1' - t_2') + \frac{u}{c^2}(x_1' - x_2')\right]}{\sqrt{1 - \frac{u^2}{c^2}}}, \tag{B.1}$$

which simplifies to

$$l = \frac{(x_2' - x_1')(1 - \frac{u^2}{c^2})}{\sqrt{1 - \frac{u^2}{c^2}}}. \tag{B.2}$$

Then, since by definition $l' = x_2' - x_1'$, we find that

$$l = \sqrt{1 - \frac{u^2}{c^2}}\, l', \tag{B.3}$$

which is the Lorentz Contraction. Also note that the primed time increment $(t_2' - t_1')$ makes absolutely no difference since it completely cancels out in the deriviation.

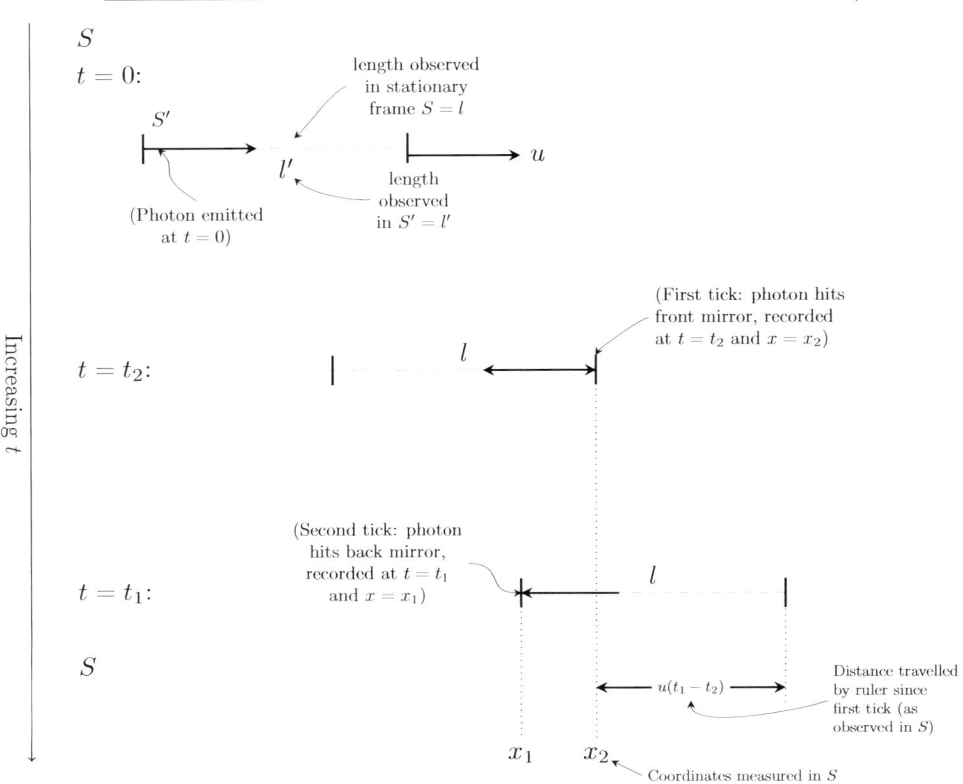

FIGURE B.1. Geometry of the photon clock. With increasing t, the clock moves to increasing x.

Calculation of the Velocity of the CM for the Case $A + B \rightarrow C + D$ $(N = 2)$

C.1. Transformation from the Lab Frame to the CM Frame

In Chapter 7,

$$\left(\vec{\mathcal{P}}, \frac{\mathcal{E}}{c} \right) \tag{C.1}$$

is a four vector. Recall too from Chapter 7 that the masses given are always rest masses. We certainly can calculate an instantaneous veclocity V between the Lab and CM frames. Consider the transformation from the Lab frame (unprimed) to the CM frame (primed), for which we obtain $\mathcal{P}' = |\vec{\mathcal{P}'}|, \mathcal{P} = |\vec{\mathcal{P}}|$

$$\mathcal{P}' = \gamma \left(\mathcal{P} - \frac{V}{c} \frac{\mathcal{E}}{c} \right), \tag{C.2}$$

where we also assume motion in the x-direction only. (See Fig. 7.1 and Eq. (4.5a).)

However, by definition

$$\mathcal{P}' = 0, \tag{C.3}$$

which implies that

$$V = \frac{\mathcal{P}c^2}{\mathcal{E}}, \tag{C.4}$$

which we evaluate initially in the Lab.

From Eqs. (7.4), (7.16), (7.3), and (7.14), we have

$$\mathcal{P} = \frac{m_1 v_1}{\sqrt{1 - v_1^2/c^2}} + 0 = \frac{m_1 v_1}{\sqrt{1 - v_1^2/c^2}} \tag{C.5}$$

$$\mathcal{E} = \frac{m_1 c^2}{\sqrt{1 - v_1^2/c^2}} + m_2 c^2, \tag{C.6}$$

which, when substituted into Eq. (C.4), gives

$$V = \frac{m_1 v_1}{m_1 + m_2 \sqrt{1 - v_1^2/c^2}}. \tag{C.7}$$

Thus, our instantaneous frame velocity is in reality constant. Since m_1, m_2 and v_2 are all constants, the frame velocity between the Lab and CM systems is appropriate for a continuous Lorentz transformation. Also, in Eq. (C.7) if we neglect the last term in the radical in the denominator $\left(\left|\frac{v_1}{c}\right| \ll 1\right)$ we recover the classical expression for the CM velocity

$$V_{cl} = \frac{m_1 v_1}{m_1 + m_2}. \tag{C.8}$$

Note too that in Eq. (C.7), if $m_1 \to 0$ and $v_1 \to c$, the equation becomes indeterminate, so that the derivation is not valid if particle #1 is a photon.

C.2. Transformation from the CM Frame to the Lab Frame (Inverse)

We saw in subsection C.1 that, for a projectile photon, that there are serious problems in transforming from the initial Lab frame to the CM frame. However, it should be possible to transform from the final CM frame to the Lab frame (since there are no photons in the final state). Thus, we now consider, for the general problem, the inverse Lorentz transformation given by

$$\mathcal{P} = \bar{\gamma}\left(\mathcal{P}' + \frac{\bar{V}}{c}\frac{\mathcal{E}'}{c}\right) = \frac{\bar{\gamma}\bar{V}}{c^2}\mathcal{E}' \tag{C.9}$$

$$\frac{\mathcal{E}}{c} = \bar{\gamma}\left(\frac{\mathcal{E}'}{c} + \frac{\bar{V}}{c}\mathcal{P}'\right) = \frac{\bar{\gamma}\mathcal{E}'}{c}, \tag{C.10}$$

which follows since $\mathcal{P}' = 0$, and

$$\bar{\gamma} = \frac{1}{\sqrt{1 - \frac{\bar{V}^2}{c^2}}}, \tag{C.11}$$

with $\bar{V}$ the frame velocity for the inverse Lorentz transformation. Note that we take a different V, namely $\bar{V}$, for the inverse transformation. Then, Eq. (C.10) gives the result

$$\bar{\gamma} = \frac{\mathcal{E}}{\mathcal{E}'}, \tag{C.12}$$

which, from Eq. (C.11) we obtain

$$\frac{\bar{V}}{c} = \sqrt{1 - \left(\frac{\mathcal{E}'}{\mathcal{E}}\right)^2}. \tag{C.13}$$

However, from Eqs. (C.9) and (C.10), it is clear too that

$$\frac{\bar{V}}{c} = \frac{\mathcal{P}c}{\mathcal{E}}, \tag{C.14}$$

which is the same as the original Lorentz transformation V in Eq. (C.4). Thus we have established that

$$\bar{V} = V, \tag{C.15}$$

which is correct for any Lorentz and inverse Lorentz transformations. Then, from Eqs. (C.13) and (C.14) we find that

$$\frac{\mathcal{P}c}{\mathcal{E}} = \sqrt{1 - \left(\frac{\mathcal{E}'}{\mathcal{E}}\right)^2}, \tag{C.16}$$

which can be reexpressed as

$$1 - \left(\frac{\mathcal{E}'}{\mathcal{E}}\right)^2 = \frac{\mathcal{P}^2 c^2}{\mathcal{E}^2}, \tag{C.17}$$

or

$$\mathcal{E}^2 - \mathcal{E}'^2 = \mathcal{P}^2 c^2.$$ (C.18)

Eq. (C.18) can then be written as (since $\mathcal{P}' = 0$)

$$\mathcal{E}'^2 = \mathcal{E}'^2 - \mathcal{P}'^2 c^2$$ (C.19)
$$= \mathcal{E}^2 - \mathcal{P}^2 c^2$$

which gives the original threshold equation (7.18) (and which can be generalized to Eq. (7.23)).

Note that Eqs. (C.13) and (C.15) give the result

$$\frac{V}{c} = \sqrt{\frac{\mathcal{E}^2 - \mathcal{E}'^2}{\mathcal{E}^2}} = \sqrt{\frac{(E_1 + m_2 c^2)^2 - (m_3 + m_4)^2 c^4}{(E_1 + m_2 c^2)^2}},$$ (C.20)

for which there is no problem if $m_1 = 0$ and

$$E_1 = pc.$$ (C.21)

We have considered the inverse transformation from the final CM system to the initial Lab system, which is allowed due to the fact that there are no photons in the final state. Then, we have found that the problem for photons in the transformation, discussed in subsection C.1, can be resolved. Thus, particle #1 can be a photon and there is agreement with reference 8, page 95.

Finally, we remark about the indeterminacy in Eq. (C.7) for V if particle #1 is a photon. There is a limit to V for the problem we are considering; namely, with particles #3 and #4 at rest in the final CM system. The limit is given by Eq. (C.20), and indeed it is a constant. However, for other Lab-CM problems, the limit may be different, or not defined.

The Relativistic Doppler Shift, Inertial Frames at an Angle with Respect to One Another

D.1. General Moving Source and Stationary Observer

See Fig. 8.6, for which we have

$$P' = \left(p'_x, p'_y, p'_z, \frac{E'}{c} \right) \tag{D.1}$$

and

$$P = \left(p_x, p_y, p_z, \frac{E}{c} \right), \tag{D.2}$$

which are related by the usual Lorentz transformation, Eqs. (4.5). Also, recall that

$$E = h\nu' = h\nu_s = p'c = |\vec{p}'|c \tag{D.3}$$

$$E = h\nu = h\nu_\circ = pc = |\vec{p}|c \tag{D.4}$$

Note too that c/ν is the *proper wave length* for an inertial system at rest.

From the geometry in Fig. 8.6, we can write down the following equations:

$$p'_x = p'\cos\alpha' = \frac{h\nu_s}{c}\cos\alpha' \tag{D.5}$$

$$p'_y = p'\sin\alpha' = \frac{h\nu_s}{c}\sin\alpha', \tag{D.6}$$

where we see (by assumption) that $p_z = p'_z = 0$, for the photon. Then, from the Lorentz transformation Eq. (4.5d), we obtain

$$\frac{E'}{c} = \gamma\left(\frac{E}{c} - \frac{u}{c}p_x \right), \tag{D.7}$$

or

$$\frac{h\nu'}{c} = \frac{h\nu_s}{c} = \gamma\left(\frac{h\nu_\circ}{c} - \frac{u}{c}\frac{h\nu_\circ}{c}\cos\alpha \right), \tag{D.8}$$

or (cancelling factor of h and c from both sides of the equation)

$$\nu_s = \nu_\circ\gamma\left(1 - \frac{u}{c}\cos\alpha \right). \tag{D.9}$$

From another Lorentz equation, Eq. (4.5a), we find that

$$p'_x = \gamma \left(p_x - \frac{u}{c} \frac{E}{c} \right),$$
(D.10)

or

$$\frac{h\nu'}{c} \cos \alpha' = \gamma \left(\frac{h\nu}{c} \cos \alpha - \frac{u}{c} \frac{h\nu}{c} \right),$$
(D.11)

which becomes

$$\nu_s \cos \alpha' = \nu_\circ \gamma \left(\cos \alpha - \frac{u}{c} \right),$$
(D.12)

or from Eq. (D.9)

$$\cos \alpha' = \frac{\left(\cos \alpha - \frac{u}{c} \right)}{\left(1 - \frac{u}{c} \cos \alpha \right)}.$$
(D.13)

After performing some elementary trigonometry, we obtain

$$\tan \alpha' = \frac{\pm \sin \alpha}{\gamma \left(\cos \alpha - \frac{u}{c} \right)}$$
(D.14)

$$\sin \alpha' = \frac{\pm \sin \alpha}{\gamma \left(1 - \frac{u}{c} \cos \alpha \right)}.$$
(D.15)

There are several special cases:

(1) See Fig. 8.1, with a source approaching an observer head on, for which we have

$$\alpha' = \alpha = 0; \; \cos \alpha = 1$$
(D.16)

$$\nu_\circ = \frac{\nu_s}{\left[\gamma \left(1 - \frac{u}{c} \right) \right]} = \sqrt{\frac{1 + \frac{u}{c}}{1 - \frac{u}{c}}} \, \nu_s,$$
(D.17)

which is Eq. (8.14).

(2) See Fig. 8.3, with a source receding in an "opposite head-on" direction from an observer, for which we have

$$\alpha' = \alpha = \pi; \; \cos \alpha' = \cos \alpha = -1$$
(D.18)

$$\nu_\circ = \frac{\nu_s}{\gamma \left(1 + \frac{u}{c} \right)} = \sqrt{\frac{1 - \frac{u}{c}}{1 + \frac{u}{c}}} \, \nu_s,$$
(D.19)

which is Eq. (8.13).

(3) In Fig. 8.6 and Eq. (D.9) and (D.13), let

$$\alpha = \pm \frac{\pi}{2} \quad \left[\alpha' = \cos^{-1} \left(-\frac{u}{c} \right) \right]$$
(D.20)

$$\nu_\circ = \nu_s \gamma^{-1} = \sqrt{\left(1 - \frac{u^2}{c^2} \right)} \, \nu_s,$$
(D.21)

which is the transverse Doppler Effect (a red shift) described in the last part of subsection 8.2.

Now, examine Fig. 8.7, for which, in Problem #3 of Chapter 8, you are asked to prove that

$$\nu_\circ = \frac{\nu_s}{\gamma\left[1 - \frac{u}{c}\cos\alpha\right]} \tag{D.22}$$

and

$$\cos\alpha' = \frac{\left(\cos\alpha - \frac{u}{c}\right)}{\left(1 - \frac{u}{c}\cos\alpha\right)}, \tag{D.23}$$

which agree with Eqs. (D.9) and (D.13), respectively. Thus, it makes no difference whether the source or the observer moves. Again, *this is indeed Relativity*!

Appendix **E**

Gauge Invariance

A general gauge transformation is given by

$$\mathcal{A} \to \mathcal{A}' = \mathcal{A} + \partial\theta, \tag{E.1}$$

where $\mathcal{A}$ and ∂ are the four vectors defined in Eqs. (9.8) and (9.14), respectively. We thus obtain for the space-like and time-like components in Eq. (E.1):

$$\vec{A} \to \vec{A}' = \vec{A} + \vec{\nabla}\theta \tag{E.2a}$$

$$\phi \to \phi' = \phi - \frac{\partial\theta}{\partial t}. \tag{E.2b}$$

where θ is an arbitrary SpaceTime function.

Substituting Eqs. (E.2) into Eqs. (9.4), we find that

$$\vec{B}' = \vec{B} \tag{E.3a}$$

$$\vec{E}' = \vec{E}, \tag{E.3b}$$

so that Maxwell's equations, given by Eqs. (9.1), do not change. If you choose the Lorentz gauge, Eq. (9.6), then the following equation must be true:

$$\vec{\nabla} \cdot \vec{A}' + \frac{1}{c^2}\frac{\partial\phi'}{\partial t} = \vec{\nabla} \cdot \vec{A} + \frac{1}{c^2}\frac{\partial\phi}{\partial t} = 0, \tag{E.4}$$

which, from Eqs. (E.2), implies the relation

$$\Box^2\theta = 0, \tag{E.5}$$

where the d'Alembertian operator, $\Box^2$, is defined by Eq. (9.11).

For a particle of charge q, interacting with the electromagnetic field, we must change the relativistic four momentum of the particle, defined in Eq. (5.22), as follows:

$$P \to P - q\mathcal{A}. \tag{E.6}$$

(See Reference 6.)

Moreover, in all modern field theories, the equations are required to be both Lorentz and gauge invariant.

*If you can't explain it simply, you don't understand it well
enough.*

Albert Einstein

Bibliography

The Bibliography is divided into four sections: The first, which is primarily for Special Relativity; the second, which gives companion references that especially emphasize General Relativity, Astrophysics, and/or Cosmology, including material on the Big Bang, Black Holes, and Superluminal Expansion; the third, which contains additional material related to the Special Appendix; and the fourth, giving books and papers for Quantum Mechanics and the Dirac equation.

Books on Special Relativity

[1] A. Beiser. *Concepts of Modern Physics.* McGraw-Hill, NY, 1963. See Chapter 1.

[2] Peter G. Bergmann. *Introduction to the Theory of Relativity.* Prentice Hall, Englewood Cliffs, NJ, 1942. See Part 1.

[3] A. Einstein. On the electrodynamics of moving bodies. In *The Principle of Relativity, a collection of original memoirs on the special and general theories of relativity.* Dover, Mineola, NY. The original paper by Einstein, translated by W. Perrett and G. B. Jeffry, was originally published in *Annalen der Physik* in 1905.

[4] R. Feynman, R. B. Leighton, and M. Sands. *The Feynman Lectures on Physics*, volume II. Addison Wesley, MA, 1989. Chapters 25 and 26.

[5] J. Foster and J. D. Nightingale. *A Short Course in General Relativity.* Springer-Verlag, NY, 1995. First edition published in 1979, Corrected fourth printing 2001. There is a nice review of special relativity in Appendix A.

[6] H. Goldstein. *Classical Mechanics.* Addison-Wesley, Cambridge, Mass, 1953. See Chapter 6.

[7] D. J. Griffiths. *Introduction to Electrodynamics.* Addison-Wesley, NJ, 1999. See Chapter 12.

[8] James B. Hartle. *Gravity, an Introduction to Einsteins General Relativity.* Addison Wesley, NY, 2003. See part I.

[9] I. S. Hughes. *Elementary Particles.* Cambridge University Press, Cambridge, MA, second edition, 1985. See Appendix A.

[10] J. D. Jackson. *Classical Electrodynamics.* John Wiley & Sons, NY, 1975. See chapters 11 and 12.

[11] I. Landau and E. Lifshitz. *The Classical Theory of Fields.* Addison-Wesley, Cambridge, MA, 1951. See Chapters 1–7.

[12] C. Moller. *The Theory of Relativity.* Oxford University Press, Oxford, 1952. See Chapters I–VII.

[13] Richard A. Mould. *Basic Relativity.* Springer-Verlag, New York, 1994. See Part I.

[14] W.K. H. Panofsky and M. Phillips. *Classical Electricity and Magnetism.* Addison-Wesley, Cambridge, MA, 1955. See Chapters 14, 15, 16, 17, and 22.

[15] D. C. Peaslee and H. Mueller. *Elements of Atomic Physics.* PrenticeHall, NY, 1955. See Chapter 8 and Section 9.4.

[16] Edwin F. Taylor and John A. Wheeler. *Spacetime Physics, Introduction to Special Relativity.* W. H, Freeman and Company, NY, second edition, 1992.

Books on General Relativity, Astrophysics and/or Cosmology

[17] Fred Adams and Greg Laughlin. *The Five Ages of the Universe, Inside the Physics of Eternity.* The Free Press, a Division of Simon and Schuster, NY, 1999.

[18] Michell Begelman and Martin Rees. *Gravity's Fatal Attraction.* W. H. Freeman and Company, NY, 1996.

[19] M. V. Berry. *Principles of Cosmology and Gravitation.* Cambridge University Press, NY, 1998.

[20] David H. Clark and Matthew D. H. Clark. *Measuring the Cosmos, How Scientists Discovered the Dimensions of the Universe.* Rutgers University Press, New Brunswick, NJ, 2004.

[21] K. Croswell. *The Universe at Midnight.* The Free Press, NY, 2001.

[22] Kitty Ferguson. *Measuring the Universe.* Walker and Company, NY, 1999.

[23] Timothy Ferris. *Coming of Age in the Milky Way.* Anchor Books, NY, 1988.

[24] Timothy Ferris. *The Whole Shebang.* Simon and Schuster, NY, 1997.

[25] Karen G. Fox. *The Big Bang Theory.* John Wiley and Sons, NY, 2002.

[26] Donald Goldsmith. *The Runaway Universe.* Perseus Books, Cambridge, MA, 2000.

[27] Brian Greene. *The Fabric of the Cosmos.* Vintage Books, NY, 2005.

[28] Alan H. Guth. *The Inflationary Universe.* Addison Wesley Publishing Company, NY, 1997.

[29] Stephen W. Hawking. *A Brief History of Time.* Bantam Books, NY, 1988.

[30] Stephen W. Hawking. *The Universe in a Nutshell.* Bantam Books, NY, 2001.

[31] Dan Hooper. *Dark Cosmos, in Search of our Universes Missing Mass and Energy.* Smithsonian Books, NY, 2006.

[32] Barrow John D. *The Origin of the Universe.* Basic Books, NY, 1994.

[33] Jonathan I. Katz. *The Biggest Bangs.* Oxford University Press, NY, 2002.

[34] Robert P. Kirshner. *The Extravagant Universe, Exploding Stars, Dark Energy, and the Accelerating Cosmos.* Princeton University Press, Princeton, NJ, 2002.

[35] Lawrence M. Kraus. *A Universe from Nothing; Why There is Something Rather Than Nothing.* Free Press, 2012.

[36] Lawrence M. Krause. *Hiding in the Mirror.* Viking Press, NY, 2005.

[37] Lawrence Krauss. *Quintessence.* Basic Books, NY, 2000.

[38] D. Linley. *The End of Physics.* BasicBooks, NY, 1993.

[39] Malcolm S. Longair. *Our Evolving Universe.* Cambridge University Press, NY, 1996.

[40] Fulvio Melia. *The Black Hole at the Center of our Galaxy.* Princeton University Press, Princeton, NJ, 2003.

[41] Arthur I Miller. *Empire of the Stars, Obsession, Friendship, and Betrayal in the Quest for Black Holes.* Houghton Mifflin Company, NY, 2005.

[42] Roger Penrose. *The Emperor's New Mind.* Oxford University Press, NY, 1989.

[43] Richard Price, editor. *The Future of Spacetime, essays by Kip S. Thorne, Igor Novikov, Timothy Ferris, and Alan Lightman.* W. W. Norton and Oxford University Press, NY, 2002.

[44] Martin Rees. *Before the Beginning.* Addison-Wesley, Reading, MA, 1997.

[45] Martin Rees. *Our Cosmic Habitat.* Princeton University Press, Princeton, NJ, 2001.

[46] Barbara Ryden. *Introduction to Cosmology.* Addison Wesley, NY, 2003.

[47] Bernard F. Schutz. *A First Course in General Relativity.* Cambridge University Press, NY, 1998.

[48] Charles Seife. *Alpha and Omega, The search for the Beginning and End of the Universe.* Viking, NY, 2003.

[49] Joseph Silk. *A Short History of the Universe.* Scientific American Library, NY, 1994.

[50] Simon Singh. *Big Bang, The Origin of the Universe.* HarperCollins, NY, 2004.

[51] Leonard Susskind. *The Black Hole Wars, My Battle with Stephen Hawking to make the World Safe for Quantum Mechanics*. Little, Brown, and Co., NY, 2008.

[52] E. J. Taylor and J. A. Wheeler. *Exploring Black Holes, Introduction to General Relativity*. Addison Wesley, NY, 2000.

[53] Kip S. Thorne. *Black Holes and Time Warps*. W. W. Norton & Company, NY, 1994.

[54] Trinh Xuan Thuan. *The Secret Melody*. Oxford University Press, NY, 1995.

[55] Trinh Xuan Thuan. *Chaos and Harmony, Perspectives on Scientific Revolutions of the Twentieth Century*. Oxford, NY, 2001.

[56] Philippe Tourrenc. *Relativity and Gravitation*. Cambridge University Press, NY, 1997.

[57] Steven Weinberg. *The First Three Minutes, A Modern View of the Origin of the Universe*. Basic Books, NY, 1977. Afterward in 1993.

Additional References for the Special Appendix

[58] Isaac Asimov. *ATOM, Journey Across the Subatomic Cosmos*. Truman Talley Books/Plume, NY, 1992.

[59] James D. Bjorken and Sidney D. Drell. *Relativistic Quantum Mechanics*. McGraw-Hill, NY, 1964.

[60] Frank Close, Michael Martin, and Christine Sutton. *The Particle Explosion*. Oxford University Press, NY, 1987.

[61] W. N. Cottingham and D. A. Greenwood. *Electricity and Magnetism*. Cambridge University Press, NY, 1991.

[62] G. D. Coughlan and J. E. Dodd. *The Ideas of Particle Physics, An Introduction for Scientists*. Cambridge University Press, NY, second edition, 1991.

[63] Sir William Cecil Dampier. *A History of Science, and its relations with Philosophy and Religion*. Cambridge university Press, NY, fourth edition, 1949.

[64] Gordon Fraser, Egil Lillestol, and Inge Sellevag. *The Search for Infinity, Solving the Mysteries of the Universe*. Reed International Books, Ltd., NY, 1995.

[65] Douglas C. Giancoli. *Physics*. Prentice Hall, New Jersey, fourth edition, 1995.

[66] Sheldon L. Glashow. *From Alchemy to Quarks*. Brooks/Cole Publishing Co., Pacific Grove, CA, 1994.

[67] David Halliday, Robert Resnick, and Jearl Walker. *Fundamentals of Physics*. John Wiley & Sons, Inc., NY, fourth edition, 1993.

[68] Lloyd Motz and Jeffeson Hane Weaver. *The Story of Physics*. Avon Boobs, NY, 1989.

[69] Gerrit L. Vercshuur. *Hidden Attraction, the Mystery and History of Magnetism*. Oxford University Press, NY, 1993.

Books and Papers about Quantum Mechanics and the Dirac Equation

[70] J. D. Bjorken and S. D. Drell. *Relativistic Quantum Mechanics*. McGraw Hill, Inc., NY, 1964. Chapters 1–5.

[71] R. W. Davies, K. T. R. Davies, P. Zory, and D. S. Nydick. Field equations for massless bosons from a Dirac-Weinberg formalism. *Am. J. of Physics*, 78:1023, 2010.

[72] R. W. Davies, K. T. R. Davies, and D. S. Nydick. Field equations for the massive vector boson from Dirac and Weinberg formalisms. *Can. J. of Physics*, 91:506, 2013.

[73] Art Hobson. There are no particles, there are only fields. *Am. J. of Physics*, 81:211–223, 2013.

[74] S. S. Schweber. *An Introduction to Relativistic Quantum Field Theory*. Row, Peterson, and Co., NY, 1961. Part I.

[75] S. S. Schweber, H. A. Bethe, and F. de Hoffman. *Mesons and Fields*, volume Vol I, Fields. Row, Peterson, and Co., NY, 1956. Part I.

[76] S. Weinberg. *Lake Views: This World and the Universe.* Belknap of Harvard U. P., Cambridge, MA, 2011.

proper length, 30
proper time, 30, 137
    differential of, 39
proper wave length, 145

QED, 1
Quantum Electrodynamics, 1
quantum information theory, 4
Quantum Mechanics, 2, 3
quasar, 105

Röntgen, Wilhelm, 3
radiation
    Cherenkov, 87
    infrared, 3
    ultraviolet, 3
radio waves, 3
red shift, 73
red-shift parameter, 75
relative velocity, 72
relativisitic kinematics, 39
Relativistic Beaming, 78–80
relativistic dynamics, 39
relativistic kinetic energy, 43
relativistic quantum theory, 109
relativity, 113
rest mass, 41, 59
rest mass energy, 42

scattering problem
    in elementary particle physics, 63
Schrödinger's cat, 113
Schrödinger, Erwin, 2
Schwarzschild bookkeeper, 115
shell observer, 115
simultaneity, 16–17
single-particle theory, 109
slowing of light near the Sun, 104
Snell's law, 88
sonic boom, 89
SpaceTime continuum, 11
special relativity, 2, 4, 9
    dynamics in, 44
    orthogonality in, 41
spectrometer, 76
speed of light, 5
standard candle, 77

stochastic cooling, 65
straight line in two dimensions, 51
superconductors, 127
superluminal expansion of clouds, 91–92
superluminal motion, 87–92
superluminal velocity, 78

Thomson, G. P., 3
threshold case, 60
threshold problem, 59–63
tidal forces, 98
time dilation, 6, 17–19
time travel, 34
total eclipse of the Sun, 103
twin paradox, 18, 23

uncertainty principle, 113
unit vector tangent to the path, 40
universe
    acceleration of, 21

velocity of a particle, 41
velocity-dependent mass, 42
Villard, Paul, 3
von Lenard, Philip, 3

wavicles, 3
white dwarf star, 101
world line, 30
    of a free particle, 54

x rays, 3

Young, Thomas, 2

46784164R00093

Made in the USA
Lexington, KY
15 November 2015